SpringerBriefs in

SpringerBriefs in Geography presents concise summaries of cutting-edge research and practical applications across the fields of physical, environmental and human geography. It publishes compact refereed monographs under the editorial supervision of an international advisory board with the aim to publish 8 to 12 weeks after acceptance. Volumes are compact, 50 to 125 pages, with a clear focus. The series covers a range of content from professional to academic such as: timely reports of state-of-the art analytical techniques, bridges between new research results, snapshots of hot and/or emerging topics, elaborated thesis, literature reviews, and in-depth case studies.

The scope of the series spans the entire field of geography, with a view to significantly advance research. The character of the series is international and multidisciplinary and will include research areas such as: GIS/cartography, remote sensing, geographical education, geospatial analysis, techniques and modeling, landscape/regional and urban planning, economic geography, housing and the built environment, and quantitative geography. Volumes in this series may analyze past, present and/or future trends, as well as their determinants and consequences. Both solicited and unsolicited manuscripts are considered for publication in this series.

SpringerBriefs in Geography will be of interest to a wide range of individuals with interests in physical, environmental and human geography as well as for researchers from allied disciplines.

More information about this series at http://www.springer.com/series/10050

Liette Vasseur · Mary J. Thornbush
Steve Plante

Adaptation to Coastal Storms in Atlantic Canada

 Springer

Liette Vasseur
Department of Biological Sciences
Brock University
St. Catharines, ON
Canada

Mary J. Thornbush
Department of Geography
Brock University
St. Catharines, ON
Canada

Steve Plante
Département de Développement Régional et Territorial
Université du Québec à Rimouski
Rimouski, QC
Canada

ISSN 2211-4165　　　　　　　　　ISSN 2211-4173　(electronic)
SpringerBriefs in Geography
ISBN 978-3-319-63491-3　　　　　ISBN 978-3-319-63492-0　(eBook)
DOI 10.1007/978-3-319-63492-0

Library of Congress Control Number: 2017946648

© The Author(s) 2018
This work is subject to copyright. All rights are reserved by the Publisher, whether the whole or part of the material is concerned, specifically the rights of translation, reprinting, reuse of illustrations, recitation, broadcasting, reproduction on microfilms or in any other physical way, and transmission or information storage and retrieval, electronic adaptation, computer software, or by similar or dissimilar methodology now known or hereafter developed.
The use of general descriptive names, registered names, trademarks, service marks, etc. in this publication does not imply, even in the absence of a specific statement, that such names are exempt from the relevant protective laws and regulations and therefore free for general use.
The publisher, the authors and the editors are safe to assume that the advice and information in this book are believed to be true and accurate at the date of publication. Neither the publisher nor the authors or the editors give a warranty, express or implied, with respect to the material contained herein or for any errors or omissions that may have been made. The publisher remains neutral with regard to jurisdictional claims in published maps and institutional affiliations.

This Springer imprint is published by Springer Nature
The registered company is Springer International Publishing AG
The registered company address is: Gewerbestrasse 11, 6330 Cham, Switzerland

Preface

In late 2010 and early January 2011, Atlantic Canada experienced a series of severe winter storms with high winds and surge that caused important damage in several coastal communities. Some people experienced flooding, while others coastal erosion. Subsequently, in 2011, a large multisite longitudinal project was initiated to (1) better understand people's experiences with storms and (2) co-construct with these communities adaptation plans for the future, thus improving their resilience to climate change. Interviews were conducted in 2011–2012 and again in 2014 as a follow-up to examine changes in perception over time. This brief seeks to present the findings from the second set of interviews in 2014 compared to the initial findings. Based on in-depth semi-structured interviews and focus groups in 10 studied communities in Québec and New Brunswick, it was discovered that people felt resilient, but at the same time vulnerable to storms and other extreme events. While they may have been involved in the longitudinal project, the lessons learned extracted from the findings show that much more remains to be done in order to ensure that communities are prepared for future environmental and climate changes. As this project used a participatory action research approach, this brief conveys the importance of integrating local actors from various sectors and their existing knowledge when developing adaptation plans and proactive coastal management strategies.

St. Catharines, Canada	Liette Vasseur
St. Catharines, Canada	Mary J. Thornbush
Rimouski, Canada	Steve Plante

Contents

1	**Introduction**...	1
	1.1 Methodological Framework in this Study...................	3
	1.2 Scope of this Brief...	4
	References..	5
2	**Coastal Communities in Atlantic Canada**......................	7
	2.1 Demographics...	8
	2.2 Discussion...	10
	References..	13
3	**Background Research**...	17
	3.1 Participatory Action Research................................	19
	References..	25
4	**Methodology**...	29
	4.1 Introduction...	29
	4.1.1 Selecting the Communities in Atlantic Canada..........	31
	4.1.2 First Series of Interviews.............................	32
	4.1.3 Interventions and PAR.................................	36
	4.1.4 Second Series of Interviews..........................	37
	4.2 Discussion...	38
	References..	39
5	**Findings from Initial Interviews**.................................	41
	5.1 Experience with 2010–2011 Major Storm Events............	42
	5.2 Psychosocial Barriers to Change.............................	42
	5.3 Discussion...	44
	5.4 Experiences and Lessons Learned from Coastal Storms...	47
	5.5 Lessons Learned and Additional Measures.................	49
	References..	50

6	**Findings from Follow-up Interviews**		55
	6.1	Introduction	55
	6.2	Results	56
		6.2.1 Knowledge of the CCC-CURA Project	58
		6.2.2 Resilience	59
		6.2.3 Natural Environment	59
	6.3	Discussion	60
	References		62
7	**Implications and Lessons Learned**		65
	7.1	Introduction	65
	7.2	Resilience to Climate Change	67
	7.3	Integrating Governance into Social-Ecological Resilience	67
	7.4	Coastal Community Resilience Planning	68
	7.5	Moving Further in Enhancing Resilience Through Ecosystem Governance	70
	7.6	Conclusion	72
	References		73
8	**Conclusions**		77
	8.1	Summary of the Overall Findings	78
	8.2	Main Take Home Messages	79
	8.3	Contributions	80
	8.4	Further Work	81
	8.5	Discussion	82
	References		86
Index			89

Chapter 1
Introduction

Abstract This chapter introduces this brief as part of an interdisciplinary framework from an integrated social-ecological systems (SES) perspective. The described project used a participatory action approach (PAR) along with a longitudinal multisite case study to help 10 Atlantic Canada communities enhance their resilience to climate change. Interviews were conducted prior to and after a series of interventions to investigate people's perceptions at individual and community levels to the 2010 winter storms in Atlantic Canada as a basis for assessing the impacts and adaptation associated with climate change. The longitudinal approach here necessitated re-interviewing actors in 2014 to examine the changes in their perceptions over time when communities are accompanied in a process of planning for adaptation. In addition to the study scope, some of the existing knowledge-base is briefly highlighted at the end as an initial contribution to the findings of the study.

Keywords Winter storms · Climate change · Adaptation · Longitudinal study · Sustainability

Climate change is having multiscalar and cumulative effects around the world. Recent scholarly attention has been directed at adaptation efforts in order to lubricate the human transition to climate-based environmental change. Natural Resources Canada (2014a) has already identified several biophysical and socio-economic impacts due to climate change and sea-level rise (Table 1.1). Atlantic Canada, more specifically, is expected to experience more storm events, increasing storm intensity, rising sea level, storm surge, coastal erosion, and flooding.

This will especially affect coastal communities and their economic activity (fisheries, trade, tourism), traditional use (collecting clams, cemetery), land-use occupation (localization of houses), and infrastructure (bridges, roads, energy facilities, etc.), making them vulnerable. Some coastal communities are already experiencing saltwater intrusion into their freshwater aquifers, causing problems with their drinking water supply (Natural Resources Canada 2014b) or degrading piped sewage and water systems faster by corrosion. This issue is likely to be exacerbated by drier summer conditions (e.g., higher evapotranspiration), leading to

Table 1.1 Summary of impacts from Natural Resources Canada (2014a)

Biophysical impacts	Socioeconomic impacts
Higher sea-surface temperatures	Increased flood risk and potential loss of life
Higher storm-surge flooding	Damage to coastal infrastructure
More extensive coastal inundation	Increased property loss
Saltwater intrusion	Increased risk of disease
Reduced sea-ice cover	Increased length of shipping season
Increased coastal erosion	Changes in renewable and subsistence resources
Loss of coastal habitat	Loss of cultural resources and values

increased pressure on water demands and affecting domestic use, agriculture, fisheries, tourism, and energy. These communities can expect increased disruptions to transportation, electricity transmission, and communications in the future. This will tend to isolate them and could affect food safety and personal security. Although both agriculture and forestry could benefit from the anticipated higher temperatures and longer growing seasons, extreme events (droughts, heavy rainfall, hail) may impact farm water usage and crop production.

Non-climatic factors, such as demographic, social, and economic trends, may limit adaptive responses. Where the adaptive capacity is circumscribed, there is potential for vulnerability. The lack in adaptive capacity due to a high level of vulnerability can be reduced through different processes or strategies that can increase adaptation, such as planning, increased awareness, better organization, etc. These strategies can help revise emergency response measures and manage the actual situation and development along the coast. However, in aged communities, which are sparsely populated and far from political powers and where education level and average annual incomes are low, such adaptation measures will be a challenge to implement.

Between December 2010 and January 2011, Atlantic Canada experienced a series of multiple weather-related events in the form of storms (Environment Canada 2013; Whitewood and Phillips 2011). These winter storms affected the coast both physically (in terms of storm surge, flooding, marine submersion, erosion, and wave action) and socially (from the household to community level). Small rural coastal communities were impounded by a multitude of storm-related impacts. This major storm event could be reflective of what is coming up in the coastal experience of climate change in Atlantic Canada. Indeed, Lemmen et al. (2016) have reported that Canadian marine coasts are highly vulnerable to an increasing number of storms. Combined with sea-level rise and a very dynamic coast, where erosion can be severe and increasing with storms, coastal communities are becoming more vulnerable than ever to climate change. These weather events, accompanied by others over the years, now attest to the variability that is possible in the Atlantic provinces and are indicative of climate change effects (Lemmen et al. 2016).

In order to better understand how communities face these extreme events and define solutions, researchers have been using various approaches. Case studies are a very common way to examine challenges in communities and define potential strategies for them in order to enhance their resilience (Doughty 2016). They can be controversial, as they are not easily generalized, but they are useful for understanding complex issues and are relevant for community-based research (Zainal 2007). Another approach that has been increasingly used in the past decade is participatory action research (PAR). This approach has the advantage to share information between communities and researchers on a more equal footing, and allows for the co-construction of solutions (MacDonald 2012). Combining these approaches is possible with the use of various tools, such as focus groups and interviews, as methods of data collection.

This brief presents the various steps and approaches that were taken to help Atlantic Canada communities enhance their resilience in the face of climate change, mainly focusing on weather extremes. By so doing, the work aims to gauge developments in climate change adaptation from a social ecological system (SES) sustainability framework. The focus of this brief is to consider the impacts as challenges or opportunities and deliberate on potential solutions as part of climate change adaptation to major storms at the coast. Experiences range from the individual to household to community levels, embracing psychological effects as well as group responses for coping. By comparing responses from 2011–2012 based on interviews taken soon after the winter storms of 2010 with a revisit of participants in 2014, it is possible to examine interim responses and track any changes associated with research initiatives and cross-temporal effects as part of a longitudinal study. This research is highly interdisciplinary to allow for the integration of various types of knowledge as well as disciplines (from physical and geographical data to social perceptions). This was necessary in order to help these communities examine solutions from different angles and not necessarily only one, such as economy.

1.1 Methodological Framework in this Study

The current study draws from various research approaches, including:

- interdisciplinary, with ecological as well as social and territorial sciences frameworks for analysis;
- inter-sectoral, with different categories of actors coming from diverse economic spheres;
- PAR as part of community outreach and endeavors;
- longitudinal study to investigate any cross-temporal effects on perception;
- case studies, focusing on small rural communities in Atlantic Canada; and
- an integrated SES approach to examine sustainability that includes environmental or ecological as well as socioeconomic considerations.

The work recognizes the importance of multiscale action (at various levels) in order to elicit change in governance and respond to climate change adaptation as well as pinpoint the role that this capacity-building process has to play. Importantly, the research considers multilevel governance (from municipalities/local service districts or LSDs to provincial to federal levels of government) and how it affects adaptation responses, including anticipation and preparation to natural hazards stemming from a changing climate. It adopts a psychosocial framework to social science research that has ecological relevance (as in the coastal zone). This is achieved by considering both individual and social responses (community). Finally, this study is holistic in that it also investigates individual perception and how it is affected by experience.

The contribution of this research stems from its multilevel approaches (individual/family/household/community; psychosocial; multilevel governance; etc.) and inter-sectoral and interdisciplinary approaches to sustainability (environmental, social, economic, and local policy). It also contributes toward another case study for climate change adaptation in Canada and assesses community vulnerability as well as coastal resilience from an integrated systems (SES) perspective. The work assesses risks associated with climatic hazards triggered by storms and does all of this from a cross-spatial/-temporal perspective.

1.2 Scope of this Brief

The research was implemented initially with interviews with members of small rural communities located across various provinces in Atlantic Canada and followed by actions and activities in each of them. This comprised communities in the Canadian provinces of Québec (QC), New Brunswick (NB), and Prince Edward Island (PEI). We focused our attention on the diversity of governance processes at local and regional scales (municipality, LSDs, nonmunicipalized community). However, because PEI was not revisited in 2014, the focus will be on the provinces of QC and NB in Chap. 6. Ten small rural communities located along the coastal zone (a multisite study) were targeted for the actions and activities in QC and NB. Interviews were held in both English and French and were conducted both singly and in couples. Details about the adopted methodology appear in Chap. 4; the communities involved are outlined in Chap. 2; and the findings are conveyed for both sets of interviews (in 2011–2012 versus 2014) in subsequent chapters (Chaps. 5 and 6).

Collected responses informing this research were based on open and semi-directed questions as well as participant observations, with added commentary and elaboration possible. Actors came from the public and economic sectors, civil society, and nonprofit organizations, and were identified following snowball sampling. The interviews were recorded and this compilation of verbal information was also available in addition to spreadsheet summaries of responses, on which the analysis was based. Various publications have been produced from the interviews and the work carried out in these communities, including publications regarding demographics. For example, from the interviews, a paper was already published as part of a special issue in Int J Environ Res Public Health (Vasseur et al. 2015). This research article examined gender-based experiences and perceptions, with implications for a gender bias in terms of action and adaptation. More specifically, men were found to be more personally prepared for the winter storms; and they were also more active in their communities. Women showed more of an emotional response, expressing fear and worry; and their actions and felt impacts were closer to home. The importance of these findings is in their indication of the significance of demographics on action and adaptation.

This brief explicitly addresses the longitudinal component of the research, comparing responses in 2011–2012 with 2014. From such a cross-temporal perspective, it is anticipated that any effects of the research will become apparent and that any change (in perception and otherwise) can be identified through the process.

References

Doughty CA (2016) Building climate change resilience through local cooperation: a Peruvian Andes case study. Reg Environ Change 16:2187–2197

Environment Canada (2013) Canada's top ten weather studies for 2010. http://www.ec.gc.ca/meteo-weather/default.asp?lang=En&n=2552BDB2-1#t1

Lemmen DS, Warren FJ, James TS, Mercer Clarke CSL (eds) (2016) Canada's Marine coasts in a changing climate. Government of Canada, Ottawa, p 274

MacDonald C (2012) Understanding participatory action research: a qualitative research methodology option. Can J Action Res 13:34–50

Natural Resources Canada (2014a) Archived—Coastal zone. http://www.nrcan.gc.ca/environment/resources/publications/impacts-adaptation/reports/assessments/2004/ch7/10207

Natural Resources Canada (2014b) Chapter 4—Atlantic Canada. http://www.nrcan.gc.ca/environment/resources/publications/impacts-adaptation/reports/assessments/2008/ch4/10341

Vasseur L, Thornbush M, Plante S (2015) Gender-based experiences and perceptions after the 2010 winter storms in Atlantic Canada. Int J Environ Res Public Health 12:12518–12529

Whitewood R, Phillips D (2011) North America: Canada. In Blunden J, Arndt DS, Baringer MO (eds) State of the Climate in 2010. B Am Meteorol Soc 92(6):S1–S266

Zainal Z (2007) Case study as a research method. J Kemanusiaan 9:6

Chapter 2
Coastal Communities in Atlantic Canada

Abstract The population sample of 10 small rural coastal communities in Atlantic Canada is described, spanning the provinces of Québec, New Brunswick, and Prince Edward Island. Interviews of 74 people (both singly and in couples) were completed in 2011–2012 soon after the 2010 winter storms and another one was also done in 2014 as part of a follow-up survey. The focus of this chapter is to highlight the characteristics of the participants in this research, including demographics, such as gender, age, education, and occupation, which are outlined. Generally, more men than women were sampled (33 men and 22 women). The age range of interviewees was 26–90 years, with women commonly 45–54 and men with an older age range of 55–65 years old. Some traditional occupations for these communities are represented by fishers and farmers. Most of the participants resided at the coast all of their lives, with only a couple relocated there within the past 5 years at the time of interview.

Keywords Rural communities · Demographics · Sex/Gender · Age · Occupation

Coastal communities around the world are sensitive to climate change, particularly to sea-level rise and storm surge that could lead to flooding and erosion, posing a risk to people, buildings, and infrastructure. Atlantic Canada is no different and small communities there are vulnerable to changes in sea level and storm action, including high waves associated with storm surge, coastal flooding, erosion, and shoreline change. In Atlantic Canada, rising mean sea level is amplified by widespread crustal subsidence (Forbes 2008). The vulnerability of coastal communities increases due to more people occupying the coastal zone and associated rising property value in the face of poor and uneven adaptation (see Forbes 2008). According to the authors, vulnerability is affected by the timing and effectiveness of adaptation as well as coping capacity. Places that are located at low elevations are especially susceptible to sea-level rise and flooding. Atlantic Canada is normally not at risk from tropical cyclones, but can be affected by storm activity linked to hurricane activity further south along the eastern coast of North America. "Killer storms," for instance, have appeared in the North Atlantic and are known to track

© The Author(s) 2018
L. Vasseur et al., *Adaptation to Coastal Storms in Atlantic Canada*,
SpringerBriefs in Geography, DOI 10.1007/978-3-319-63492-0_2

closer to the coast, where sea-surface temperature has increased (Scheibling and Lauzon-Guay 2010). In addition, in Atlantic Canada, winter storms are more frequently impacting coastal communities due to the lack of sea surface ice, which tends to protect against high waves in the winter. Damage can be more important than during the hurricane season, as blocks of ice with waves can really damage buildings and infrastructure along the coast. It is known that vulnerability to tropical storm impacts is also affected by socioeconomic variables, as for instance inequitable resource distribution that can be overcome through income diversification, reduced poverty, common property, and collective security (Kelly and Adger 2000). This means that an integrated approach (social, economic, and environmental) is required to scrutinize climate change adaptation and sustainability.

Small rural communities situated at the coast of Atlantic Canada need to be studied in terms of their vulnerability and response to climatic change affecting the coastline. In this study, participants were interviewed either singly or in couples in their spoken language of either English or French. Based on interviews in 2011–2012, and subsequently in 2014, their experiences of coastal storms and the 2010 winter storms, in particular, were examined through the use of various questions that will be outlined in later chapters. Not all participants responded to all questions, however, so that the sample size varied.

Because their responses are the basis for the findings of this study, it is important to take into consideration the individual characteristics of the participants. Demographics, for instance, will be considered and outlined next in this chapter. However, it is worth noting that demographics do not exclusively affect individual outlook and individual/social behavior; and other variables should also be considered as well as religious beliefs (faith in an afterlife or divine intervention), which may bolster a sense of responsibility and instill a need for action (Hope and Jones 2014). Indeed, it has been suggested (Lyle 2015) that a more targeted approach to climate change adaptation involves knowledge of factors, circumstances, and historical factors affecting decision making and hierarchical systems in this process. In particular, the author brought to light the role of individual beliefs and risk perceptions as being central to such a hierarchy.

2.1 Demographics

Participants in the provinces of Québec (QC) and New Brunswick (NB) reported being significantly affected by the 2010 winter storms. By comparison, respondents in Prince Edward Island (PEI), especially women, responded more often "No" to whether they had been affected by the 2010 winter storms (Table 2.1). Women in PEI were particularly unaffected by the 2010 storms. For this reason, the focus will be on the other provinces (QC and NB).

The majority of participants were professionals as well as in male-dominated occupations, such as primary producers and the trades (Table 2.2). Women had a similar range of occupations than men, with a total of 59 occupations listed for

2.1 Demographics

Table 2.1 Summary of (%) participants affected by the 2010 storms

Overall (N = 67)	Female	Male	Couples	Total
Yes	13.4	32.8	4.5	50.7
No	16.4	29.8	3.0	49.2
Total	29.8	62.6	7.5	99.9
QC (n = 17)	Female	Male	Couples	Total
Yes	23.5	41.2	0.0	64.7
No	5.9	23.5	5.9	35.3
Total	29.4	64.7	5.9	100.0
PEI (n = 10)	Female	Male	Couples	Total
Yes	0.0	20.0	0.0	20.0
No	50.0	30.0	0.0	80.0
Total	50.0	50.0	0.0	100.0
NB (n = 63)	Female	Male	Couples	Total
Yes	12.7	34.9	4.8	52.4
No	9.5	31.7	6.3	47.5
Total	22.2	66.6	11.1	99.9

Table 2.2 Occupational codes of participants (counts) affected by storms

Province	Female	Male	Couples
Overall	Business/financial (5) Professional—biologist (3) Education (3) Public servant (2) Professional—environment/conservation Professional—food engineering Professional—health Professional—psychology Professional—social worker Trade	Public servant (9) Trade (7) Primary producer (7) Business/financial (5) Professional—environment/conservation (3) Education (2) Professional—biologist Professional—computers Professional—environmental science	Education (2) Primary producer Trade
QC	Business/financial (2) Education Trade	Public Servant (3) Education Professional—environmental science Business/financial Trade Primary producer Professional—environment/conservation	

(continued)

Table 2.2 (continued)

Province	Female	Male	Couples
PEI	Professional—biologist Professional—Environment/conservation Business/financial Public servant	Professional—biologist Trade Professional—computers Professional—environment/conservation Primary producer	
NB	Professional—biologist (2) Business/financial (2) Education (2) Professional—food engineering Professional—health Professional—psychology Public servant Professional—social Worker	Public servant (6) Primary producer (5) Trade (5) Business/financial (4) Professional—environment/conservation Education	Education (2) Primary producer Trade

those participants who had experienced storms. Most participants held volunteer/community positions and some (15%) were retired. At the regional level, the territory appeared to be multifunctional in terms of economy.

2.2 Discussion

The age range of participants who had previous experience with storms was 29–90 years. For females, the most common age range was slightly lower (45–54) than for men (55–64). In Canada, Atlantic Canada has the oldest population (Vasseur and Catto 2008), with the majority of residents occupying coastal communities (Rapaport et al. 2015). Age is an important determinant of response for various reasons, including different modes of media accessed by the elderly versus students, for instance, to obtain information about severe storms (Silver and Conrad 2010). They may, for instance, check media to access advisories and information regarding storm characteristics in order to assess their own risk (Dow and Cutter 1998). Elderly people are vulnerable to climate-related hazards, and this will increase with projected aging populations in developed countries around the world as well as aging infrastructure on which we depend (Oven et al. 2012). In countries like South Korea, for example, older age groups are more vulnerable to meteorological disasters, including typhoons (Myung and Jang 2011).

Similarly, other factors make groups vulnerable, such as social and cultural discrimination as part of the "sociocultural dimension" (Lambrou and Nelson 2013), as for instance affecting youth (Overton 2014). Due to a lack of education or employment opportunities, isolated coastal communities often experience an "exodus" or out-migration of youth. This situation exacerbates the conditions, where

2.2 Discussion

less tax revenues limit the capacity of the communities to develop adaptation plans and better infrastructure. In such circumstances, elderly people are also leaving in order to be closer to healthcare centers. Marginalized groups, including women and the poor (Denton 2002) as well as children/the aged, disabled, and indigenous people, etc. (Dominelli 2013), are missing opportunities to prepare for climate change impacts and have the least capacity for representation and negotiation (Demetriades and Esplen 2008). It is, therefore, important for their social networks to be considered when developing strategies to overcome barriers (Messias et al. 2012). Those with a circumscribed adaptive capacity are more vulnerable to the shocks and stresses associated with climate change impacts (Polack 2008). Rather than depending on an elite group for their survival (Wong 2009), the poor should have improved access to resources in order to be able to respond effectively on their own. Women are similarly affected due to their roles (family and work) and access to resources (Alston 2013). In Nepal, for example, inter-caste dependencies as well as gender inequalities are affected by adaptation processes (Onta and Resurreccion 2011). In rural India, their ability to access resources affects women's vulnerability (Roy and Venema 2002; espoused by Tschakert 2012 as part of political ecology). Even in our communities, most women tended to be at home or working in more precarious conditions than men (Vasseur et al. 2015).

Occupation is a vital consideration because of its impact on income and poverty. It has also been linked to the preferred adaptation strategies used by men and women during floods and droughts (Codjoe et al. 2012; Tatlonghari and Paris 2013). Typically gender-specific occupations, such as fishing (Weeratunge et al. 2010), farming, and charcoal production (cf. Codjoe et al. 2012), can limit adaptation. It is not only their occupation that affects women's adaptation, as increased female employment (Godden 2013; and whether or not they are employed or paid for their roles, cf. Vasseur 2016) can determine their adaptation to climate change impacts. Women have been found to be more vulnerable than men because of their income and education (cf. Enete 2013), which, in turn, according to Liu et al. (2014), affect their knowledge and perceptions of climate change. Gender, and not age, has been found to significantly shape risk perception (Safi et al. 2012); however, others have attributed gender and education as well as age as affecting risk perception in NB (e.g., Lieske et al. 2014). Research has conveyed the influence of age on opinions regarding climate variation, but other variables have also included religion and educational level (Teka et al. 2013).

Examples from around the world portray women's vulnerability to climate change. Women have suffered more flood risks (e.g., sexual harassment) than men in Vietnam (Tu and Nitivattananon 2011). Household income and gender are known to affect concern over the societal impacts of heat waves in Adelaide, Australia (Akompab et al. 2013). In Australia, there is increasing gender-based violence in cases of drought, for example, where there is income-related stress (Whittenbury 2013). However, more studies are needed that examine local adaptive strategies (Djoudi and Brockhaus 2011; Reed et al. 2014; Sultana 2014 for multiscalar research investigating gender and social relations). To illustrate this idea, Ketlhoilwe (2013) reports that the main adaptive strategies employed by women include the integration of local knowledge as well as new technology and social interaction leading to social learning. Women's networks and their capacity-building determine the impacts of environmental change on them (Makhabane 2002). Their role to maintain community cohesion has been observed (e.g., Vasseur 2016), even though they may not receive any direct income for performing these roles. Where women are isolated, as for example confined women in Bangladesh; and they are more affected by climate change impacts, which poverty and health also impound (Shabib and Khan 2014).

Women are generally more trusting of face-to-face communication and where information is received from people that they know (Alber 2013). This makes sense, especially in light of experiences associated with Hurricane Katrina, where data convey nationally did not always reflect local experiences (Anthony and Sellnow 2011). Data communicated through various streams, including radio, television, and the Internet are considered to be trusted sources of information, and can be deployed to enhance preparedness and education as well as inform resilience strategies (Burger 2015). Information acquired by women are from diverse sources and quality; for example, it could be information that they attained through their multiple roles in the household and communities that can be used to warn of impending disasters and prepare for them (Ross-Sheriff 2007).

The manuscript by Vasseur et al. (2015) represents one of the few published studies to investigate the effects of gender on experiences and perceptions of winter storms in Atlantic Canada. The authors have conveyed the demographics already outlined above; in addition, they revealed some aspects of gender-based experiences and perceptions. For example, both men and women experienced similar impacts as a result of storms, particularly flooding (see their Table 2, p. 12523). However, women tended to refer to flooding at home, whereas men mentioned flooding of roads and infrastructure. Women also mentioned damage to their personal belongings. The gendered experience was also affected by sex-typical occupations, as for instance male fishers and farmers. This latter point affected the male response to erosion, with eroded farmland affecting farmers.

The gender-based division of labor that is evident in these small rural communities points to gender roles affecting the experiences and perceptions of men and women in Atlantic Canada. In addition, it is also crucial to examine age-related impacts, especially as 15% were retired in this study sample. Age can affect mobility and adaptability, which can affect evacuation and responses to storm-related effects and hazards. These aging communities will respond in more traditional ways and convey patterns associated with traditional gender roles, as for instance women responding closer to home rather than in the overarching social domain over immediate community connections. Evacuation is likely to become more important as a climate adaptation strategy for vulnerable communities (Kuhl et al. 2014), but elderly people may be unable to do so and, therefore, would be more at risk.

Problems arising from income-based limitations set by occupation and employment status (that could also be affected by gender as well as age demographics and education), could slow down recovery as well as trigger higher impacts and affect risk; as for example for low-income women, who are at greater risk than others (cf. Ajibade et al. 2013) and, thereby, more vulnerable and less resilient to flooding. Likewise, fishers with more fishing-gear investment and government support can adapt occupationally as well as through networks (family, fishers' groups, etc.); conversely, their adaptation (in, e.g., Mozambique) is hindered by limited assets and declining resources, competition, and poverty (Blythe et al. 2014). In addition, the diversification of occupation (or livelihood diversification, Goulden et al. 2013), along with an improved economy and sense of community (or "community coherence"), all act to reduce vulnerability and enhance resilience, as to storm surge in Iceland (Geirsdóttir et al. 2014).

References

Ajibade I, McBean G, Bezner-Kerr R (2013) Urban flooding in Lagos, Nigeria: patterns of vulnerability and resilience among women. Glob Environ Chang 23:1714–1725

Akompab DA, Bi P, Williams S, Grant J, Walker IA, Augoustinos M (2013) Awareness and attitudes towards heat waves within the context of climate change among a cohort of residents in Adelaide, Australia. Int J Environ Res Public Health 10:1–17

Alber G (2013) Gendered access to green power: motivations and barriers to changing energy provided (chapter 10). In: Alston M, Whittenbury K (eds) Research, action and policy: addressing the gendered impacts of climate change. Springer, Berlin, pp 135–148

Alston M (2013) Introducing gender and climate change: research, policy and action (chapter 1). In: Alston M, Whittenbury K (eds) Research, action and policy: addressing the gendered impacts of climate change. Springer, Berlin, pp 3–16

Anthony KE, Sellnow TL (2011) Information acquisition, perception, preference, and convergence by Gulf Coast residents in the aftermath of the Hurricane Katrina crisis. Argumentation Advocacy 48:81–96

Blythe JL, Murray G, Flaherty M (2014) Strengthening threatened communities through adaptation: insights from coastal Mozambique. Ecol Soc 19(2):6

Burger J (2015) Ecological concerns following Superstorm *Sandy*: stressor level and recreational activity levels affect perceptions of ecosystem. Urban Ecosyst 18:553–575

Codjoe SNA, Atidoh LK, Burkett V (2012) Gender and occupational perspectives on adaptation to climate extremes in the Afram Plains of Ghana. Clim Change 110:431–454

Demetriades J, Esplen E (2008) The gender dimensions of poverty and climate change adaptation. IDS Bull 39(4):24–31

Denton F (2002) Climate change vulnerability, impacts, and adaptation: why does gender matter? Gender Dev 10(2):10–20

Djoudi H, Brockhaus M (2011) Is adaptation to climate change gender neutral? Lessons from communities dependent on livestock and forests in northern Mali. Int For Rev 13(2):123–135

Dominelli L (2013) Gendering climate change: implications for debates, policies and practices (chapter 6). In: Alston M, Whittenbury K (eds) Research, action and policy: addressing the gendered impacts of climate change. Springer, Berlin, pp 77–94

Dow K, Cutter SL (1998) Crying wolf: repeat responses to hurricane evacuation orders. Coast Manage 26:237–252

Enete AA (2013) Challenges of agricultural adaptation to climate change: the case of cassava post-harvest in southeast Nigeria. Int J Clim Change Strateg Manage 5(4):455–470

Forbes DL (2008) Climate-change impacts and adaptation: a coastal geoscience perspective. Atlantic Geol 44:11

Geirsdóttir GE, Gísladóttir G, Jónsdóttir Á (2014) Coping with storm surges on the Icelandic South Coast: a case study of the Stokkseyri village. Ocean Coast Manage 94:44–55

Godden NJ (2013) Gender and declining fisheries in Lobitos, Perú: beyond *Pescador* and *Ama De Casa* (chapter 18). In: Alston M, Whittenbury K (eds) Research, action and policy: addressing the gendered impacts of climate change. Springer, Berlin, pp 251–264

Goulden MC, Adger WN, Allison EH, Conway D (2013) Limits to resilience from livelihood diversification and social capital in lake social-ecological systems. Ann Assoc Am Geogr 103(4):906–924

Hope ALB, Jones CR (2014) The impact of religious faith on attitudes to environmental issues and Carbon Capture and Storage (CCS) technologies: a mixed methods study. Technol Soc 38:48–59

Kelly PM, Adger WN (2000) Theory and practice in assessing vulnerability to climate change and facilitating adaptation. Clim Change 47:325–352

Ketlhoilwe MJ (2013) Improving resilience to protect women against adverse effects of climate change. Clim Dev 5(2):153–159

Kuhl L, Kirshen PH, Ruth M, Douglas EM (2014) Evacuation as a climate adaptation strategy for environmental justice communities. Clim Change 127:493–504

Lambrou Y, Nelson S (2013) Gender issues in climate change adaptation: farmers' food security in Andhra Pradesh (chapter 14). In: Alston M, Whittenbury K (eds) Research, action and policy: addressing the gendered impacts of climate change. Springer, Berlin, pp 189–206

Lieske DJ, Wade T, Roness LA (2014) Climate change awareness and strategies for communicating the risk of coastal flooding: a Canadian Maritime case example. Estuar Coast S Sci 140:83–94

Liu Z, Smith WJ Jr, Safi AS (2014) Rancher and farmer perceptions of climate change in Nevada, USA. Clim Change 122:313–327

Lyle G (2015) Understanding the nested, multi-scale, spatial and hierarchical nature of future climate change adaptation decision making in agricultural regions: a narrative literature review. J Rural Stud 37:38–49

Makhabane T (2002) Promoting the role of women in sustainable energy development in Africa: networking and capacity-building. Gender Dev 10(2):84–91

Messias DKH, Barrington C, Lacy E (2012) Latino social network dynamics and the Hurricane Katrina disaster. Disasters 36(1):101–121

Myung H-N, Jang JY (2011) Causes of death and demographic characteristics of victims of meteorological disasters in Korea from 1990 to 2008. Environ Health 10:82

Onta N, Resurreccion BP (2011) The role of gender and caste in climate adaptation strategies in Nepal. Mt Res Dev 31(4):351–356

Oven KJ, Curtis SE, Reaney S, Riva M, Stewart MG, Ohlemüller R, Dunn CE, Nodwell S, Dominelli L, Holden R (2012) Climate change and health and social care: defining future hazards, vulnerability and risk for infrastructure systems supporting older people's health care in England. Appl Geogr 33:16–24

Overton LR-A (2014) From vulnerability to resilience: an exploration of gender performance on art and how it has enabled young women's empowerment in post-hurricane New Orleans. Procedia Econ Financ 18:214–221

Polack E (2008) A right to adaptation: securing the participation of marginalized groups. IDS Bull 39(4):16–23

Rapaport E, Manuel P, Krawchenko T, Keefe J (2015) How can aging communities adapt to coastal climate change? Planning for both social and place vulnerability. Can Public Pol 41(2):166–177

Reed MG, Scott A, Natcher D, Johnston M (2014) Linking gender, climate change, adaptive capacity, and forest-based communities in Canada. Can J For Res 44:995–1004

Ross-Sheriff F (2007) Women and disasters: reflections on the anniversary of Katrina and Rita. J Women Soc Work 22(1):5–8

Roy M, Venema HD (2002) Reducing risk and vulnerability to climate change in India: the capabilities approach. Gender Dev 10(2):78–83

Safi AS, Smith WJ Jr, Liu Z (2012) Rural Nevada and climate change: vulnerability, beliefs, and risk perception. Risk Anal 32(6):1041–1059

References

Scheibling RE, Lauzon-Guay J-S (2010) Killer storms: North Atlantic hurricanes and disease outbreaks in sea urchins. Limnol Oceanogr 55(6):2331–2338

Shabib D, Khan S (2014) Gender-sensitive adaptation policy-making in Bangladesh: status and ways forward for improved mainstreaming. Clim Dev 6(4):329–335

Silver A, Conrad C (2010) Public perception of and response to severe weather warnings in Nova Scotia, Canada. Meteorol Appl 17:173–179

Sultana F (2014) Gendering climate change: geographical insights. Prof Geogr 66(3):372–381

Tatlonghari GT, Paris TR (2013) Gendered adaptations to climate change: a case study from the Philippines (chapter 17). In: Alston M, Whittenbury K (eds) Research, action and policy: addressing the gendered impacts of climate change. Springer, Berlin, pp 237–250

Teka O, Houessou GL, Oumorou M, Vogt J, Sinsin B (2013) An assessment of climate variation risks on agricultural production: perceptions and adaptation options in Benin. Int J Clim Change Strateg Manage 5(2):166–180

Tschakert P (2012) From impacts to embodied experiences: tracing political ecology in climate change research. Geogr Tidsskrift-Danish J Geogr 112(2):144–158

Tu TT, Nitivattananon V (2011) Adaptation to flood risks in Ho Chi Minh City, Vietnam. Int J Clim Change Strateg Manage 3(1):61–73

Vasseur L (2016) Diversifying the garden: a way to ensure food security and women empowerment (chapter 7—Burkina Faso). In: Fletcher AJ, Kubik W (eds) Women in agriculture worldwide: key issues and practical approaches. Gower Publishing, Surrey, pp 103–115

Vasseur L, Catto N (2008) Atlantic Canada In Lemmen DS, Warren FJ, Lacroix J, Bush E (eds) From impacts to adaptation: Canada in a changing climate 2007. Government of Canada, Ottawa, pp 119–170

Vasseur L, Thornbush M, Plante S (2015) Gender-based experiences and perceptions after the 2010 winter storms in Atlantic Canada. Int J Environ Res Public Health 12:12518–12529

Weeratunge N, Snyder KA, Sze CP (2010) Gleaner, fisher, trader, processor: understanding gendered employment in fisheries and aquaculture. Fish Fish 11:405–420

Whittenbury K (2013) Climate change, women's health, wellbeing and experiences of gender based violence in Australia (Chapter 15). In: Alston M, Whittenbury K (eds) Research, action and policy: addressing the gendered impacts of climate change. Springer, Berlin, pp 207–222

Wong S (2009) Climate change and sustainable technology: re-linking poverty, gender, and governance. Gender Dev 17(1):95–108

Chapter 3
Background Research

Abstract The main research objective of this study and the approaches used to collect the data are outlined in this chapter. In particular, it focuses on the participatory action research (PAR) and multiple site (multisite) approaches used in this project. It also conveys the longitudinal aspect of the project. The long-term perspective adopted here serves from a sustainability standpoint to inform planning and policy. In particular, one of the greatest contributions so far in the published literature from this project has been that of gender-based adaptation. Using demographics, the consideration of gender roles and experiences as well as specific impacts and responses, it has been possible to examine how men and women are both affected and responded to the winter storms and the implications for gender mainstreaming in climate change adaptation research.

Keywords Multiple site (multisite) approach · Participatory action research (PAR) approach · Longitudinal (cross-temporal) study · Resilience · Sustainability

The main goal of the interview component of the project was to assess the perceptions of participants located in small rural coastal communities situated in the study provinces of Québec (QC), New Brunswick (NB), and Prince Edward Island (PEI) to their experiences of the 2010 winter storms (and other storms in general). Experiences were considered at various levels, from individual/household to community and regional (province-based). Demographics were noted for the participants in order to understand the population sample and to inform their responses (see previous chapter). Furthermore, a list of questions was used to identify their experiences and understandings of the situation that they experienced. The first set of interviews (2011–2012) and the second set in 2014 were different, except for some questions, and are, therefore, considered in separate chapters. This means that the longitudinal views were based on specific questions related to their experiences with storms. It is also important to note that PEI was not included in the 2014 series of interviews, as no further activities with these communities were conducted after the first set of interviews. The advantage of having several communities was that the multisite approach enabled for multilevel comparison, from local to regional.

© The Authors(s) 2018
L. Vasseur et al., *Adaptation to Coastal Storms in Atlantic Canada*,
SpringerBriefs in Geography, DOI 10.1007/978-3-319-63492-0_3

Temporality may not be obvious, but remains relevant, as revealed by longitudinal studies (e.g., Ford et al. 2013), such as those spanning 43 years (1965–2008, Munji et al. 2014) or even longer. As expressed by these authors, locally relevant perceptions of individuals and their responses allow for the identification of the needs and issues (challenges as well as opportunities) or adversities affecting resilience and decision making as part of adaptive management and co-management that is multisite and scale-specific and affects learning through experience (Plummer and Armitage 2007). Furthermore, the dynamics of "community resilience" over time (Amundsen 2012; "community resilience" encompasses several variables: structural design, knowledge of risks, prevention and warning, governance, and recovery, Ewing et al. 2010) is another facet of the research that needs to be considered and developed.

By comparing the impacts and effects of various storms, and not just the 2010 winter storms, it is possible to piece together a juxtaposition of information that could serve to inform the response and, ultimately, governance processes of these communities. In particular, possessing reliable information can serve individual-to-household and community preparedness and reduce public health risk as well as improve resilience. This was true for Superstorm Sandy, which made landfall in New Jersey on 29 October 2012 (Burger and Gochfeld 2015). Here, resident's actions dealt primarily with preparedness (getting homes ready for the storm, acquiring generators and emergency supplies, sooner evacuation, better warnings, etc.) rather than to improve resilience or recovery, which should encompass governmental actions, such as restricting beach-front homes, better building standards, dune restoration, and so on (Burger and Gochfeld 2014a). Burger and Gochfeld (2014b) have suggested stronger evacuation enforcement, better preparedness information, preparation for prolonged power outages (e.g. the acquisition of generators), and attention to medical needs in order to resolve issues of health and property impacts stemming from Superstorm Sandy and future storms.

While the various disciplinary (theoretical) models of resilience have been studied extensively in the literature (e.g. Brown and Westaway 2011; Folke 2006), what has not been examined to the same extent is whether, at the local level, community members have a good understanding of the term "resilience" and its implications for themselves and their communities. Fewer studies have examined the factors that influence how people interpret resilience applied within their own context. Vasseur et al. (submitted), for instance, examined the perceptions of resilience and what it meant for participants. Part of this chapter summarizes the key findings of this manuscript as it is important for better understanding the context in which this study was initiated. It specifically explores the research questions related to how they defined resilience and what factors affect them. These questions are both relevant and important to investigate further as community action and lower level governance become more prominent as part of grassroots movements that are encompassed in (bottom-up) approaches that are currently shaping policy implementation through decision making and adaptive planning initiatives (Urwin and Jordan 2008).

In the study communities, we discovered that the word "resilience" was not often understood and, in most cases, only linked to the individual (Vasseur et al. submitted). The results demonstrated that resilience is viewed in a contextual basis by people and in this case mostly linked to coping with changes or storms. This was important to understand initially in order to determine the level of actions and where the needs were felt the most in these communities. How resilience is defined (or understood) affects its measurement and use in planning and policy, making this a considerable field-verification of the term.

Establishing the notion of resilience within sustainability science involves clarification of the concept descriptively as well as analytically and supported by case studies in order to allow it to be operationalized and applied (Brand and Jax 2007). This indicates that, in addition to analytical studies, relevant research should also encompass descriptive approaches to investigate resilience. So that what could seem as a limitation of the current study as descriptive and qualitative, could make a viable contribution to understanding and responding to change from a conceptual basis that is necessarily based on a bottom-up approach. The importance of a locally-informed and possibly activated (grassroots) decision-making process cannot be overstated, particularly at a time when local collective (Karlsson and Hovelsrud 2015) or community action is gaining interest and momentum in planning and policy-setting (Amaratunga 2014).

Sustainability adopts an integrated socioeconomic and environmental framework that is, likewise, relevant to this work. In particular, such a holistic approach (cf. Bailey and Wilson 2009) allows for consideration of the various effects, as for instance of extreme events associated with climate change, such as the storms experienced in Atlantic Canada. There are consequences also for the economy, and these can provide a basis for stress felt by the study participants. There are also environmental impacts that have implications for society and whether people choose to remain living in these communities or relocate (even just seasonally) in order to avoid being impounded by coastal storms. Finally, sustainability adopts a long-term perspective that is necessary in this kind of work and links with the complexity of the social ecological system (SES) involved.

3.1 Participatory Action Research

Since vulnerability is affected by adaptive capacity, it is both relevant and important to consider a variety of approaches (theoretical and methodological) to policy making. For instance, proactive approaches to intervention are preferred because they build adaptive capacity over more reactive policy approaches (Budreau and McBean 2007). According to these authors, adaptation can be affected by any level of government to facilitate practice and reduce vulnerability, which is "determined by a community's collective capacity to adapt to changes to the environment"

(p. 1307). Adaptive capacity is affected by factors including: education, skills, wealth, technology, infrastructure, available information, access to resources, and management capabilities. The proactive approach involves foresight, as both short- and long-term goals are kept in scope. Reactive strategies, on the other hand, are used retrospectively when an event occurs and, although it can build a short-term coping capacity, this approach normally does not help to build adaptive capacity. For example, risk assessment and management as policy tools have the potential to build adaptive capacity. In the case of the cod fishery, government intervention came too late, as collapse had already occurred in 1992 when something should have been done in the mid-1980s; there was already evidence in the 1970s of a pending collapse and proactive political strategies could have been utilized then in order to avoid the fishery collapse (Budreau and McBean 2007). According to the authors, several lessons could have been learned, as for instance surrounding the misconception that funding will automatically facilitate adaptation. In this case, adaptation also had a cultural element and historic relationship among actors that the reactive approach did not recognize. Foresight and planning in this case did not help to build the adaptive capacity required for effective adaptation, probably because of the lack of recognition of the cultural traditions that necessitated a people-participatory process.

With a participatory approach, it is possible to evoke both economically and culturally sensitive solutions that take into consideration value systems, including cultural values (McIntyre-Tamwoy et al. 2013). This is a relevant approach to Canada as a nation, as it recognizes its cultural diversity and range of value systems. Proactive policy enables social, political, economic, and environmental aspects to be recognized and may actually be more cost-effective in the long term (Budreau and McBean 2007). Consultation has been suggested by these authors with stakeholders, involving landowners, business people, etc., in order to assess the probability of risks affecting a broad spatial range and considering social, cultural, and economic implications. Proactive interventions, in particular, are necessary when cultural change is slow in order to ensure that foresight and policy are combined for economic development to occur at the community to regional scale. Donner and Webber (2014), for instance, suggest embracing a culturally appropriate planning approach spanning some 20 years of "short-term" planning without ignoring longer term temporal spans. Planning based on a participatory approach also requires local inputs, as from local governments (Broto et al. 2015).

As noted by Lindeman et al. (2015) as regards sea-level adaptation planning, "in many communities, planning has formally begun" (p. 557) along the American Atlantic coast. Using text analysis based on breakout groups in east Florida, North Carolina, and Massachusetts, these authors found (based on a 75% response rate) that professional stakeholders (including academics and representatives from federal or state agencies) are most represented in meetings involving adaptation-planning workshops. Information needs identified during these sessions include analytic and predictive tools as well as communicative and policy tools for effective planning. They recognized the importance of webinars for cost-effective information transfer as well as the need for people-centric messages.

The development and the use of "common language" can counteract misunderstandings and translation challenges between scientists and non-professionals. According to Baird et al. (2014), a PAR approach, like any other learning-oriented approach, can support climate change adaptation because of its ability to bring together a diversity of actors cross-sectorally, so that it can enhance adaptive capacity and resilience. It can work to integrate through vertical and horizontal connections as it undergoes cycles of inquiry, reflection, and action (Parkes and Panelli 2001). As a methodology, PAR encourages social learning (through engagement) and can be used to work toward adaptive management (Mackenzie et al. 2012). In an SES perspective, PAR is particularly interesting because it allows for co-construction within both social and ecological dynamics.

The transfer of knowledge is affected by participation (general understanding of the public concerning environmental issues) in coastal communities and their experience of the past, including past storms in 2000 and 2010 (Chouinard et al. 2015). For example, erosion is noted to extend some 10 m since 1975 (p. 15) and 0.5 m over 6 years (p. 16), and is thought to be linked to more frequent winter freeze-thaw episodes affecting physical weathering. Regional variations have also been noted, with extreme events in the southeast coast of NB being not as aggressive (p. 16). Residents expressed anxiety during winter storms, with stress alleviated when protection measures were adopted. They were willing to take advice from community authorities or groups and this tended to reduce feelings of frustration and tension over land-use conflicts and erosion-protection measures. Collective action, or "community solutions," has been seen as necessary for the development of community attachment. On one hand, this is weakened by an influx of wealthy retirees and summer residents (staying only 6 weeks in the year). While, on the other hand, population out-migration can affect the collective capacity due to limited number of active people. The increasing population at the coast means that more people will be affected by the acceleration of storms observed to occur since 20 years ago. There is an overwhelming demand for policies and regulations to counteract uncontrolled development, with one of the main challenges being opening dialogue to reach a community vision and planning for climate change adaptation that considers the socioeconomic and demographic composition of the region. These problems may only get worse in the southern Gulf of the St. Lawrence, where accelerated sea-level rise, which is expected alongside increasing storm intensity and diminished winter ice, leading to more coastal erosion and shoreline retreat driving future hazards (Forbes et al. 2004).

According to Chouinard et al. (2015), small community actions have long-term impacts. This is especially the case in terms of learning as well as mitigating negative impacts, reducing hazard exposure, and maintaining land and infrastructure. However, residents feel that close collaboration with local authorities and a greater cooperation of the provincial government with communities is both necessary and important. Their perspectives convey that "communities had progressed from awareness to action" (p. 21). In their study, common objectives encompass land protection and the reduction of winter storm impacts. Cooperation can also be enhanced when emergency plans counteract community isolation. There is building of mutual aid at various levels, including in rural communities. However, a lack of funding remains a restriction in some cases, as for instance for the building of Cocagne Bridge, whereas individual protective measures were on the rise. Associations were thought to promote the dissemination of information and act as social facilitators in part of a "democratic body."

One of the obstacles affecting rural coastal communities is that they are anchored in traditionalism, such as fishing dependency as employment, restricting their capacity to innovate and, therefore, find new solutions to adapt to climate change (Beaudin 1996). The moving out of youth, particularly educated youth, and now also the elders from these communities is contributing to an aging population and the exodus of new economic activities. Behavioral change that could foster a rapid changing of mentalities is difficult in the instance of youth out-migration and aged communities, as for example is evident in fishing communities (Hovelsrud et al. 2013). Beaudin (1996) also emphasizes that these communities and villages are not a homogenous whole, but instead represent a mosaic that is regionally disparate and may require a bottom-up approach. In addition, an integrative science approach can benefit assessments of behavioral responses (and change) and partake of fisheries governance (Miller et al. 2010), for instance, affecting policy options and resilience. Behavioral change is capable of enhancing resilience, as evidenced by fishers' responses after experiencing hurricanes to remove fish pots during hurricane months and mooring boats (Forster et al. 2014).

Induced behavioral change can alter demands for services, the provision of charts, and ice routing and breaking by Fisheries and Oceans Canada (Shackell et al. 2014). Small-craft harbor personnel interviews by Shackell et al. (2014) in 2012 indicate problems with (outdated) infrastructure in the midst of storm surge and rapid ice breakup as well as high-tide levels, as in the Bay of Fundy. For example, site-specific vulnerability and action prioritization for aged infrastructure in need of repair and affected by winter freeze-thaw cycles are recommended. The consideration of ice as both protector and assailant on coastal infrastructure has to be included in any planning. Priorities are not the same everywhere, however, and in some cases structural issues (damage to buildings and infrastructure) can be of a lower concern than nonstructural vulnerability issues associated with earthquake and tsunami hazards, including inadequate hazard awareness, communication, and response logistics (Wood et al. 2002).

In order to properly assess what is needed for behavioral change to occur, individuals need to be interviewed with a PAR approach. This approach is notably effective in situations where there is low social cohesion affecting decision making, so that individual (rather than collective) action is more realistic or where there are weak inter-community bonds impeding support for organizations (cf. Barnett and Eakin 2015). According to Hurlimann et al. (2014), collective action that occurs within communities (or intracommunity) should also operate between communities (at the inter-community level). In some cases, community engagement through project participation (as in interviews, focus groups, MEGF, SWOT, or workshops) can stimulate action (Cone et al. 2013) at various levels. Locally, it is possible to use workshops as venues where to establish community adaptation priorities (cf. Picketts et al. 2013). These local-scale participation efforts can translate to resilient SESs when implemented as part of multilevel management (Lovecraft and Meek 2011). According to McNeeley (2012), strategic collective action can foster "sustainable adaptation."

For communities, consideration needs to be given to the structure or governance of decision making, which can be gauged through PAR. For instance, to access the "cultural dimension" of adaptation, communities are a key point for investigation in order to determine any inequalities in rank and hierarchical governance systems that may be in place (cf. Codjoe and Issah 2016). In addition, whether communities perceive themselves to be a single unit or socially coherent affects their response, as for example whether they aid each other. These decisions, which are affected by the fabric of communities and the way that they are executed in the decision-making process, influence adaptation choices and action.

Because individual demographics affect perception, as for instance according to gender, age, and education (Deressa et al. 2011; Molua 2009), it is necessary to consider individual-to-household level responses to climate change adaptation in addition to socioeconomic capacity. It is noteworthy that lower level decisions are affected by higher up political influences, and sustainable governance needs to consider both aspects for resilient coping strategies to emerge. Akter et al. (2016) recently examined gender effects on preferences among farmers for weather-indexed crop insurance and have shown that women have significant insurance aversion (stemming from distrust and skepticism due to previous experiences with financial fraud). In this case, it is crucial that individual views are gauged so that demographics are considered at the individual-to-household level. This study found 75% of women not to be highly active outside the household domain and to possess lower levels of education, making them entrust their financial decisions to male household members. For this reason, individual interviews are necessary (rather than at the household scale, which might be more represented by men) in order to access the feminine voice.

Women's household roles tend to be more household-centered, whereas men occupy more public roles, such as liaising with government administrators and performing outdoor activities and plans (Lane and McNaught 2009). According to these authors, men make most decisions concerning resource allocation after disasters. The authors noted that, as observed by Stephanie Zoll in 2008 (in Lane and McNaught 2009; also others, i.e. Boetto and McKinnon 2013), these roles reflect employment patterns in the community, with women being confined to the village and men going further afield to periurban areas. This also determines the gender-based access to information. Participatory approaches, which tend to involve bottom-up experts and decision makers (rather than top-down approaches, e.g. at the national scale, typically upheld by scientists and policymakers, Gray et al. 2014), together with gender analysis, are empowering tools for community adaptation research because they permit communities to drive issues and processes, as we have seen in our own project (Vasseur et al. 2015).

As noted by Plante et al. (2016), community actors are often not invited to participate in finding strategies and solutions for adaptation and to improve resilience. According to the authors, top-down approaches are adopted to decision making, as evident in the governance of two coastal communities in QC (namely, in Maria and Bonaventure, Baie des Chaleurs), where learning about their own SESs helped to link up people and the environment and discuss potential solutions that are more relevant to the communities. Here, adopting a PAR approach was found to be instrumental to building resilience through improved governance. Specifically, community resilience planning was achieved through the Method of Evaluation by Group Facilitation (MEGF) used to improve governance and resilience. Such a PAR approach is inclusive, developing solutions in partnership with various actors and stakeholders, including researchers who can evoke the participation of different actors (elected officials, public and economic sectors, civil society, and nonprofit organizations). According to these authors, PAR can be reached through various public engagement techniques (e.g., MEGF, participative mapping, kitchen assemblies, focus groups, interviews, etc.). They relayed the outputs, which have included:

- community resilience plans;
- improved understanding of the consequences of hazards and risks;
- identification of adaptation solutions; and
- monitoring and evaluation of actions.

An important contribution of such a PAR approach is the greater awareness of issues and perceptual change, with long-lasting community effects.

More attention is needed to address "on-the-ground" interpretations of resilience by community members (Vasseur et al. submitted), such as sociopsychosocial effects as well as individual psychological concerns. These may impound on perceptions that affect the ability of households to cope with shock or "resilience" (Schwarz et al. 2011). It has been discovered by Vasileiadou and Botzen (2014), for instance, that individuals with previous experience of an intense, life-threatening event express more concern regarding extreme events, so that referring to personal experiences and emotions (rather than professional and second-hand experiences) should be employed when framing adaptation measures and/or seeking societal support for these, this can be considered part of an "adaptation motivation" (cf. Grothmann et al. 2013). Such a "psychological adaptation" approach considers intraindividual to social processes that may be affecting risk perception and coping responses as well as any behavioral change (Reser and Swim 2011). Other authors have recognized the effect of place attachment as a subjective response involving values, culture, and place, where people are motivated by their emotional connection to the place where they live, as for instance has been evident in northern Norway (Amundsen 2015).

References

Akter S, Krupnik TJ, Rossi F, Khanam F (2016) The influence of gender and product design on farmers' preferences for weather-indexed crop insurance. Glob Environ Change 38:217–229

Amaratunga CA (2014) Building community disaster resilience through a virtual community of practice (VCOP). Int J Disaster Resilience Built Environ 5(1):66–78

Amundsen H (2012) Illusions of resilience? An analysis of community responses to change in northern Norway. Ecol Soc 17(4):46

Amundsen H (2015) Place attachment as a driver of adaptation in coastal communities in Northern Norway. Local Environ 20(3):257–276

Bailey I, Wilson GA (2009) Theorising transitional pathways in response to climate change: technocentrism, ecocentrism, and the carbon economy. Environ Plann A 41:2324–2341

Baird J, Plummer R, Pickering K (2014) Priming the governance system for climate change adaptation: the application of a social-ecological inventory to engage actors in Niagara, Canada. Ecol Soc 19(1):3

Barnett AJ, Eakin HC (2015) "We and us, not I and me": justice, social capital, and household vulnerability in a Nova Scotia fishery. Appl Geogr 59:107–116

Beaudin M (1996) The marginalization of coastal areas: the fishing communities of the St. Lawrence Gulf. J Reg Sci 19(2):161–174

Boetto H, McKinnon J (2013) Rural women and climate change: a gender-inclusive perspective. Aust Soc Work 66:234–247

Brand FS, Jax K (2007) Focusing the meaning(s) of resilience: resilience as a descriptive concept and a boundary object. Ecol Soc 12(1):23

Broto VC, Boyd E, Ensor J (2015) Participatory urban planning for climate change adaptation in coastal cities: lessons from a pilot experience in Maputo, Mozambique. Curr Opin Environ Sustain 13:11–18

Brown K, Westaway E (2011) Agency, capacity, and resilience to environmental change: lessons from human development, well-being, and disasters. Ann Rev Environ Res 36:321–342

Budreau D, McBean G (2007) Climate change, adaptive capacity and policy direction in the Canadian North: can we learning anything from the collapse of the east coast cod fishery? Mitig Adapt Strateg Glob Change 12:1305–1320

Burger J, Gochfeld M (2014a) Perceptions of personal and governmental actions to improve responses to disasters such as Superstorm Sandy. Environ Hazards 13(3):200–210

Burger J, Gochfeld M (2014b) Health concerns and perceptions of central and coastal New Jersey residents in the 100 days following Superstorm *Sandy*. Sci Total Environ 481:611–618

Burger J, Gochfeld M (2015) Concerns and perceptions immediately following Superstorm Sandy: ratings for property damage were higher than for health issues. J Risk Res 18(2):249–265

Chouinard O, Rabeniaina TR, Weissenberger S (2015) Transfer of knowledge and mutual learning on the Canadian Atlantic coast. Chapter 2 in Coastal Zones, Elsevier, pp 13–25

Codjoe SNA, Issah AD (2016) Cultural dimension and adaptation to floods in a coastal settlement and a savannah community in Ghana. GeoJournal 81:615–624

Cone J, Rowe S, Borberg J, Stancioff E, Doore B, Grant K (2013) Reframing engagement methods for climate change adaptation. Coast Manage 41:345–360

Deressa TT, Hassan RM, Ringler C (2011) Perception and adaptation to climate change by farmers in the Nile basin of Ethiopia. J Agr Sci 149:23–31

Donner SD, Webber S (2014) Obstacles to climate change adaptation decisions: a case study of sea-level rise and coastal protection measures in Kiribati. Sustain Sci 9:331–345

Ewing L, Flick RE, Synolakis CE (2010) A review of coastal community vulnerabilities toward resilience benefits from disaster reduction measures. Environ Hazards 9:222–232

Folke C (2006) Resilience: the emergence of a perspective for social-ecological systems analyses. Glob Environ Change 16:253–267

Forbes DL, Parkes GS, Manson GK, Ketch LA (2004) Storms and shoreline retreat in the southern Gulf of St. Lawrence. Mar Geol 210:169–204

Ford JD, McDowell G, Shirley J, Pitre M, Siewierski R, Gough W, Duerden F, Pearce T, Adams P, Statham S (2013) The dynamic multiscale nature of climate change vulnerability: an Inuit harvesting example. Ann Assoc Am Geogr 103(5):1193–1211

Forster J, Lake JR, Watkinson AR, Gill JA (2014) Marine dependent livelihoods and resilience to environmental change: a case study of Anguilla. Mar Policy 45:204–212

Gray SRJ, Gagnon AS, Gray SA, O'Dwyer B, O'Mahony C, Muir D, Devoy RJN, Falaleeva M, Gault J (2014) Are coastal managers detecting the problem? Assessing stakeholder perception of climate vulnerability using fuzzy cognitive mapping. Ocean Coast Manage 94:74–89

Grothmann T, Grecksch K, Winges M, Siebenhüner B (2013) Assessing institutional capabilities to adapt to climate change: integrating psychological dimensions in the adaptive capacity wheel. Nat Hazards Earth Syst Sci 13:3369–3384

Hovelsrud GK, West J, Dannevig H (2013) Fisheries, resource management and climate change: local perspectives of change in coastal communities in Northern Norway (chapter 12). In: Sygna L, O'Brien K, Wolf J (eds) A changing environment for human security: transformative approaches to research, policy and action. Routledge, pp 135–146

Hurlimann A, Barnett J, Fincher R, Osbaldiston N, Mortreux C, Graham S (2014) Urban planning and sustainable adaptation to sea-level rise. Landscape Urban Plan 126:84–93

Karlsson M, Hovelsrud GK (2015) Local collective action: adaptation to coastal erosion in the Monkey River Village, Belize. Global Environ Chang 32:96–107

Lane R, McNaught R (2009) Building gendered approaches to adaptation in the Pacific. Gender Dev 17(1):67–80

Lindeman KC, Dame LE, Avenarius CB, Horton BP, Donnelly JP, Corbett DR, Kemp AC, Lane P, Mann ME, Peltier WR (2015) Science needs for sea-level adaptation planning: comparisons among three U.S. Atlantic coastal regions. Coast Manage 43:555–574

Lovecraft AL, Meek CL (2011) The human dimensions of marine mammal management in a time of rapid change: comparing policies in Canada, Finland and the United States. Mar Policy 35:427–429

Mackenzie J, Tan P-L, Hoverman S, Baldwin C (2012) The value and limitations of participatory action research methodology. J Hydrol 474:11–21

McIntyre-Tamwoy S, Fuary M, Buhrich A (2013) Understanding climate, adapting to change: indigenous cultural values and climate change impacts in North Queensland. Local Environ 18(1):91–109

McNeeley SM (2012) Examining barriers and opportunities for sustainable adaptation to climate change in Interior Alaska. Clim Change 111:835–857

Miller K, Charles A, Barange M, Brander K, Gallucci VF, Gasalla MA, Khan A, Munro G, Murtugudde R, Ommer RE, Perry RI (2010) Climate change, uncertainty, and resilient fisheries: institutional responses through integrative science. Prog Oceanogr 87:338–346

Molua EL (2009) Accommodation of climate change in coastal areas of Cameroon: selection of household-level protection options. Mitig Adapt Strateg Glob Change 14:721–735

Munji CA, Bele MY, Idinoba ME, Sonwa DJ (2014) Floods and mangrove forests, friends or foes? Perceptions of relationships and risks in Cameroon coastal mangroves. Estuar Coast Shelf S 140:67–75

Parkes M, Panelli R (2001) Integrating catchment ecosystems and community health: the value of participatory action research. Ecosyst Health 7(2):85–106

Picketts IM, Curry J, Déry SJ, Cohen SJ (2013) Learning with practitioners: climate change adaptation priorities in a Canadian community. Clim Change 118:321–337

Plante S, Vasseur L, Da Cunha C (2016) Adaptation to climate change and Participatory Action Research (PAR): lessons from municipalities in Quebec, Canada. In: Knieling Jörg (ed) Climate adaptation governance in cities and regions. Theoretical fundamentals and practical evidence. Wiley, London, pp 69–88

Plummer R, Armitage D (2007) A resilience-based framework for evaluating adaptive co-management: linking ecology, economics and society in a complex world. Ecol Econ 61:62–74

References

Reser JP, Swim JK (2011) Adapting to and coping with the threat and impacts of climate change. Am Psychol 66(4):277–289

Schwarz A-M, Béné C, Bennett G, Boso D, Hilly Z, Paul C, Posala R, Sibiti S, Andrew N (2011) Vulnerability and resilience of remote rural communities to shocks and global changes: empirical analysis from Solomon Islands. Global Environ Chang 21:1128–1140

Shackell N, Greenan BJW, Pépin P (2014) Climate change impacts, vulnerabilities and opportunities analysis of the Marine Atlantic Basin. Canada Ocean and Ecosystem Sciences Division; Canada, Department of Fisheries and Oceans, pp 329–379

Urwin K, Jordan A (2008) Does public policy support or undermine climate change adaptation? Exploring policy interplay across different scales of governance. Glob Environ Chang 18:180–191

Vasileiadou E, Botzen WJW (2014) Communicating adaptation with emotions: the role of intense experiences in raising concern about extreme weather. Ecol Soc 19(2):36

Vasseur L, Thornbush M, Plante S (2015) Gender-based experiences and perceptions after the 2010 winter storms in Atlantic Canada. Int J Environ Res Public Health 12:12518–12529

Vasseur L, Plante S, Znajda, SK, Thornbush M (submitted) How coastal community members perceive resilience: a case from Canada's Atlantic Coast. Ambio

Wood NJ, Good JW, Goodwin RF (2002) Vulnerability assessment of a port and harbor community to earthquake and tsunami hazards: integrating technical expert and stakeholder input. Nat Hazards Rev 3:148–157

Chapter 4
Methodology

Abstract In this chapter, we describe the methodological approach that was selected in order to better understand how communities affected by interventions on climate change adaptation change over time and whether these interventions were effective or not. The project was based on two main elements: a longitudinal study and participatory action research (PAR). The main goal of the longitudinal component of the project was to examine changes over time, while PAR aimed to co-produce knowledge and co-construct solutions with the communities. This was important in order to consider existing knowledge with scientific knowledge for more socially acceptable solutions. To do so, different tools were employed in the communities, depending on their interests and the issues that the communities wanted to work on in priority.

Keywords Longitudinal study · Participative action research (PAR) · Interventions · Interviews · Public engagement

4.1 Introduction

Different methodological approaches and tools have been used in the past to better understand how communities can enhance their adaptive capacity in the face of climate change in order to improve their resilience and engage them in this path (Smith 2016). Plenty of adaptation planning guides exist and vary in complexity depending on where and for whom they are developed (Smith 2016). Mangoyana et al. (2012) have extracted some of the principles that may be important to keep in mind when such projects or case studies are being implemented. For example, adaptation should be executed in such a way that sustainable development is maintained in the community. This would include economic activities, environmental sustainability (maintenance of ecosystem functions and services), social aspects, including livelihoods, traditions, equity, etc., and governance (Mangoyana et al. 2012). Governance is considered to be important in order to ensure that communities can work together, especially when facing risks and hazards.

© The Author(s) 2018
L. Vasseur et al., *Adaptation to Coastal Storms in Atlantic Canada*,
SpringerBriefs in Geography, DOI 10.1007/978-3-319-63492-0_4

One of the cornerstones for all adaptation planning processes is to ensure that communities are involved. While policies at governmental levels can be seen as useful, they may not always be effective due to limited enforcement. Frameworks are in place in some countries that begin to integrate climate change adaptation into long-lived projects and investments (e.g. hydroelectric plants, Mangoyana et al. 2012). As stated by Sherman and Ford (2013, p. 418): "Engaging stakeholders in assessing vulnerability and implementing adaptation interventions, however, is widely regarded to be an important factor for enhancing adaptation implementation and success."

In this study, we used participatory action research (PAR) to ensure that actors in the communities were involved from the start in the development of the process and, at the same time, examined the role that experts could play in helping communities develop adaptation strategies to climate risks. Participatory action research has been used in climate change adaptation in order to improve the integration of local existing (cultural, ecological, or traditional) knowledge into decisions. It also means that researchers often learn as much from local people than the knowledge that they bring to them. For example, Mapfumo et al. (2013) research on PAR combines with "field-based farmer learning approaches" (p. 10). This approach leads to the co-production of knowledge and co-construction of solutions with the community, since the process is iterative and research activities are directly influenced by the results coming from the interactions (Fals-Borda 1987). Participatory action research, in our case, was based on three principles, as defined by Cargo and Mercer (2008): (1) importance to include existing knowledge in defining solutions; (2) ensure that all people, especially the vulnerable, are included for social equity; and (3) make sure that all actors can define and determine the solutions, as they are the ones who will have to deal with the consequences.

When this project started, researchers debated about the influence of an event, especially an extreme event, in the rate of adapting to climate change (Vasseur 2011). However, Mangoyana et al. (2012) have been more affirmative and have suggested that "research has also demonstrated that individuals and communities are much more likely to respond to experiences of current climate variability, such as a recent flood or damage from a hurricane, than to expected or future climactic change" (p. 13). It is based on the premise that having been recently affected makes people more prone to adapt than those who see this potentially in the future. While this may be true and has been observed (e.g. McSweeney and Coomes 2011), this remains to be well documented.

The Coastal Community Challenges-Community-University Research Alliance (CCC-CURA) project was well-positioned as a series of severe storms occurred at the beginning of the project. Several communities had been affected in Atlantic Canada, while others not. To understand if being affected by storms truly enhanced the willingness of people and communities to adapt faster to climate change, we used a comparative case study approach with communities affected and not-affected. We have also referred to this approach as a longitudinal study as well,

4.1 Introduction

since communities were followed through activities for a period of over 2 years. This qualitative approach had several advantages in our case, since it allowed us to better generalize the trends in those communities (Yin 2003). In this chapter, we summarize the various research activities and tools that were used in order to extract the information and data that are presented in the rest of the chapters. They include the initial interviews, the resilience planning kit used with most communities to develop adaptation strategies, the various tools that were used to engage people in communities, and finally the last interviews that were completed in order to examine whether the interventions that were done by the research teams and whether being affected or not by 2010 storms really influenced their actions. We first start by explaining how the communities were selected.

4.1.1 Selecting the Communities in Atlantic Canada

The first step in this project was to determine which communities would be selected as case studies. Most of Atlantic Canada is comprised of small rural coastal communities and political/governmental systems vary significantly within and among provinces. The provinces targeted in our project were initially Québec (QC), New Brunswick (NB), and Prince Edward Island (PEI). However, due to changes in the research team in PEI, only the first interviews were completed and no further activities were performed under this project. The criteria that were used for the selection of the communities were that all communities had to have less than 9999 inhabitants according to the Statistics Canada (2011) census. As a second criterion, we wanted to include different types of administrative units, such as incorporated municipalities, non-incorporated municipalities known as "local service districts (LSDs)" in NB, regional municipalities, etc. The third criterion was that no recent major local, governmental, or research initiatives on climate change adaptation had been completed in these communities in order to assess the impacts that the CCC-CURA project was going to have. The final criterion was that the municipal councils or the local authorities of these selected communities endorsed the approach and were interested to be a case study. Members of the research team approached the potential communities and proposed to the group which ones would be selected. For each community, a prediagnostic profile was completed and included the information described in Table 4.1. The information can be found on the website of the CCC-CURA project (http://aruc.robvq.qc.ca/en/bibliotheque/aruc). By the end, 10 communities were selected. These communities were distributed in five ecological zones in the Estuary (Ste.-Flavie) and Gulf (Rivière-au-Tonnerre) of the St. Lawrence, the Baie des Chaleurs (Maria and Bonaventure), the Acadian Peninsula (Ste.-Marie-St.-Raphaël and Shippagan), the Northumberland Strait in the south of the Gulf (Cocagne, Grande-Digue, and Dundas), and PEI (Stratford and Morell). Their brief description can be found in Table 4.2.

Table 4.1 List of information items that were collected in each selected community prior to initiating the case study

Economic data	Economic sectors, average annual income, employment rate, etc.
Geographic data	Distribution and proportion of population living at the coast, proportion and location of cottages, location of major urban center infrastructures, road accessibility, land use plans, presence of protection and support structures and authorities for their maintenance, etc.
Administrative information	Local government system and non-municipal characteristics, presence and number of elected or non-elected councilors, relevant land use planning and municipal by-laws, presence of provincial or federal elected representative in the region, etc.
Demographics	Population size and density, population history, average age of the population, level of poverty, percentage of illiteracy, percentage of retired people, education level, emergency services Main stakeholders, number of permanent residences, presence of citizens or community groups and other cultural and social activities, languages spoken in community, existence of conflicts/tensions average property price, proportion of permanent residences along the coast, etc.
Environmental data	Geology, presence of salt marshes, dunes, cliffs, forests, agricultural lands, national or provincial protected areas, the presence of endangered species, water quality and quantity, waste-water treatment, etc.
Climate and biophysical data	History of storms, floods, heat waves, melting and ice conditions, coastal vulnerability or risk maps, frequency of extreme weather events, erosion rate, etc.
Major environmental issues	Reported in local media and facing people in the community. Duration, types of issues, solutions, level of publicity

4.1.2 First Series of Interviews

As part of the initial baseline data collection, semi-structured interviews were conducted (after research ethics approval at multiple universities) in the winter and spring of 2012. To reach a representative range of people in the communities, including the public and economic sectors, civil society, and nongovernmental organizations or NGOs, participants were initially recruited by a process of personal and public invitations. Additional participants were then added using snowball sampling, where a participant may suggest another person who can also be interviewed. A total of 74 interviews were carried out (see Table 4.2). The interviews lasted between 40 and 75 min. They were all recorded and transcribed for further analysis, such as coding by NVivo v.10.

Six different themes were included in this first series of interviews. They were: (1) experience with storms; (2) financial capital; (3) social capital; (4) vision for the future; (5) information sources; and (6) understanding of what is resilience. The first theme, experience with storms, examined how participants perceived the recent

4.1 Introduction

Table 4.2 Brief description of the 10 selected communities, including whether or not they were affected by the 2010 storms, population size, and key characteristics

Community	2010 storm damage	Pop'n 2011[†]	Key features[†]
QC: Gulf of St. Lawrence Estuary (11 interviews)			
Ste.-Flavie	Yes—surge, flooding, erosion	919	Francophone. Economic activities include tourism, agriculture and commerce. The majority of residents work outside of Ste.-Flavie in the broader geographical district
Rivière-au-Tonnerre	No	307	Francophone. Strong history in fisheries; crab fishing is still main activity today. Tourism on the rise with lengthening seasons associated with climate change
QC: Baie des Chaleurs (7 interviews)			
Maria	Yes—flooding, surge, evacuations	2536	Francophone. Settled by Acadians in mid-1700s. Twice as many residents have a university degree compared to provincial average. The hospital employs 45% of the population. Sandy habitat without the protection of rock cliffs
Bonaventure	No	1017	Francophone. Settled in 1700 s by Acadians. Economic areas include government, health, services, and finance in addition to agriculture, forestry, fishing, and tourism. Rock cliffs provide some protection from storm surge
NB: Baie des Chaleurs/Acadian Peninsula (28 interviews)			
Ste.-Marie-St.-Raphael	Yes—erosion, flooding	955	Includes the LSDs of Cap-Bateau and Pigeon Hill on the Island of Lamèque. Francophone. Lies within the Acadian Peninsula District Planning Commission (CAPA). Economic activities include agriculture, manufacturing, peat collection, fishing, and tourism. Highly exposed to storm events due to location on an island at the end of a peninsula jutting into the Atlantic Ocean
Shippagan	No	2603	Francophone. Large proportion of population employed by campuses of the University of Moncton and the New Brunswick Community College as well as the Aquarium and Marine Centre of NB. Also fishing, aquaculture, and tourism
NB: Northumberland Strait (19 interviews)			

(continued)

Table 4.2 (continued)

Community	2010 storm damage	Pop'n 2011[†]	Key features[†]
Cocagne-Grande Digue	Yes	2317 + 2182	Francophone population. LSD, meaning most services (fire, water, waste, etc.) administered by the province. Close to a large urban center (Moncton; Dieppe) fostering tourism and seasonal residences
Dundas	Not as much	6282	Francophone population. LSD. Meaning most services (fire, water, waste, etc.) administered by the province. Economy relies on fishing, agriculture and tourism. Located further inland than Cocagne-Grande Digue, and less affected by storms
Region 4—Pei (10 interviews)			
Morell	Yes—erosion	313	Anglophone population. Includes several small unincorporated villages and a segment of Abegweit First Nation's territory. Many residents commute to larger centers for employment. Location on north shore makes it highly vulnerable to northerly storm surge
Stratford	No	8574	Anglophone population. Located 5 km across the Hillsborough River Bridge from the provincial capital of Charlottetown, where many residents work. Also agriculture, tourism, recreation, and oyster fishing. On a peninsula, surrounded by water

The number of people interviewed per region is also noted

storms if they had experienced them or other storms in the past. Questions related to psychological (e.g., anxiety, depression, separation, family problems), social (e.g., community help, organization, conflicts with neighbors), and economic (e.g., damage to infrastructure, loss of land, economic activities, or employment) consequences were asked. Different aspects of governance, such as community support and decision making, were also discussed with the participants. For example, we asked questions as to whether or not they had received help from neighbors or emergency services following the storms and if they felt that they would have liked to have more support. These questions targeted them directly or what they also perceived in the community. They were also asked if they thought that they personally and their community were prepared for these extreme weather events. Other questions related to what they had learned from the storms and the actions that they intended to take to be more prepared.

4.1 Introduction

The second theme was related to the financial burdens that storms may have on participants. Indeed, Bennett et al. (2016) argue that coastal communities tend to be highly vulnerable, as they rely on natural-resource exploitation and the global market can fluctuate rapidly. This means that these fluctuations can greatly affect disposable income that may be needed for adaptation. For communities, limited financial resources may also impact the type and capacity of emergency services they can offer. We, therefore, asked questions about whether they thought they had sufficient means and their capacity to adapt personally or if the finances of the community were strong enough to enhance their level of preparedness.

Social capital was analyzed in the third theme of the interview. Social capital can be considered as a "strong predictive power in effective disaster response and recovery at the community and individual levels" (Reininger et al. 2013, p. 51). In our study, we asked whether storms had affected their relations with neighbors, family members, or the rest of the community. Questions around issues such as tensions and conflicts in the community as well as who helped with recovery were asked. This also allowed us to evaluate their level of preparedness and capacity to adapt through their decisions and values that they attached to their experiences with extreme weather events.

Understanding how interviewees envisioned the future personally and for the community in the face of climate change and experiences with storms was the fourth theme examined. To do so, we asked them what could change in the community and challenges they believed they or the community would face over the next 5–10 years when experiencing more events. It was also a way to find out the likelihood that they would experience more storms in the future and their capacity to cope with them.

The fifth theme related to the sources of information that they were using to be informed concerning the advent of extreme weather events, climate change, and how to adapt to climate change. Access to information may be critical for people and their communities. The respondents were asked in what context they were looking at information, the sources of information (e.g. Internet, newspapers, article, radio), and their appreciation of the quality of information. This can help them to better assess their situation and make proper decisions regarding the types of adaptation strategies they can use (Schechtman and Brady 2013).

The perception and how they defined resilience was the sixth theme of the interviews. In this case, we wanted to understand what resilience meant and how they perceive their own resilience or that of their community. Resilience can be defined in different ways in regards to climate change (Vasseur et al. submitted). Various models of resilience exist and have led to such differences in definitions and understanding of the term. For example, some people may relate more to social resilience (Adger 2003), for example, while others to social-ecological resilience (Carpenter et al. 2001). In addition, they wanted to know if they believe that they, their community, and the natural environment were resilient and how they related to terms. Finally, demographic questions, such as age, education, gender, and employment, were asked. It is known that demographics can greatly affect the capacity and level of preparedness of people in the face of extreme weather events (Reininger et al. 2013).

4.1.3 Interventions and PAR

Stakeholder engagement is a crucial step in any planning for adaptation to climate change (Fazey et al. 2007). However, it is clear that effective stakeholder engagement can be challenging (Sherman and Ford 2013). To enhance the chance that interventions can be successful, each community involved had first the support of its local government. Then, through the interviews and public announcements, it was possible to attract people to the meetings and activities. It is important to note that each region had a research sub-team that was dedicated to the activities in their own regions. Interventions started in 2012, after the completion of the interviews, and greatly varied among regions. This depended on the issues that the communities had to face and the type of local government they had. For example, Cocagne was an LSD and, therefore, adaptation planning was more complicated, as there was no municipal council to endorse an adaptation plan. In this case, the interventions were more with community groups than in municipalities with elected members.

The tool that was used in most communities to enhance resilience and define potential adaptation strategies was developed by Vasseur (2012) called *A Kit to Implement Dialogue on Planning Community Resilience to Environmental and Climate Changes*. The objective of this kit was to encourage and help communities implement a community dialogue in order to develop a consensus on which community elements are vulnerable to environmental and climate changes and from there develop plans that focus on strengthening the resilience of the community. It is a step-by-step explanation of the method from assessing vulnerability and gradually moving toward defining the potential solutions or strategies to adapt to climate change and, thus, enhance resilience (see Fig. 4.1). It is a training tool or a guide for community leaders and NGOs. In order to go through the kit, there is a need to engage the community. The process of encouraging people to come and help define solutions can be challenging. To do so, several types of activities were used in the various regions.

In QC, the researchers used the visions, actions, and partnerships (VAP) tool, Method of Group Evaluation Facilitation (MGEF), the monitoring of commitments and expectations tool (SWOT), among others. For example, VAP is a participatory tool that can often be used to first understand from where participants are coming in terms of climate change in their communities and then define the elements that can contribute to the success of developing an adaptation plan. It is also a tool with which people can participate in developing monitoring and evaluation plans for the future (Beaulieu et al. 2002, 2016). The MGEF is a tool that helps to define priority issues. This participatory approach enhances the ownership of participants in what is needed to move forward. Through discussion, participants list the issues that they believe are important and then vote using ballots on the first priority issue that they would like to tackle. As expected, these issues varied among communities. These techniques helped to co-construct solutions that could be implemented to move forward with climate change adaptation at the community level (e.g. inspired by SWOT, the Strategic Plan of Ste.-Flavie, which was constructed with the support of

4.1 Introduction

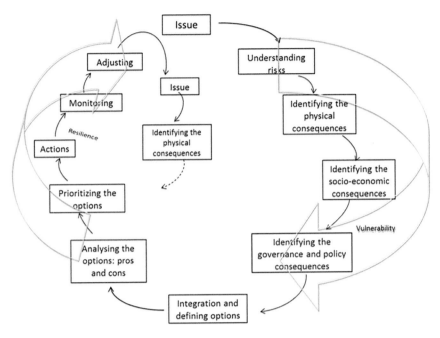

Fig. 4.1 Step-by-step process used in the planning kit used in the communities that were involved in the CCC-CURA project

the Municipalité Régionale de Comté of Mitis. Other public sessions, for example on future projections under climate change scenarios, were also organized to increase people's awareness of the subject.

4.1.4 Second Series of Interviews

Following over 2 years of interventions in the communities, a second series of interviews were completed in order to examine the changes in perceptions of the people in these communities. A total of 38 interviews were carried out (note that none were completed in PEI) and lasted between 40 and 75 min (Table 4.3). In this second series of interviews, five focus groups were added to help confirm what people stated individually. As described in the first series of interviews, audio recordings were transcribed and coded using NVivo v.10.

In the interviews, we essentially addressed the same topic as in the first series of interviews. However, a few additional questions were included on the natural environment and the role of public health and emergencies services. A final question related to their participation and perception of the CCC-CURA project.

Table 4.3 Number of interviews and focus groups carried out in the various regions during the second series of interviews

Place	2014 (Phase 2)
Northumberland Strait in the south of the Gulf (Cocagne, Grande-Digue, and Dundas)	5 more + 1 focus group
Acadian Peninsula (Ste.-Marie-St.-Raphaël and Shippagan)	10
Baie des Chaleur (Maria and Bonaventure)	11 + 2 focus groups
Gulf (Rivière-au-Tonnerre) of the St. Lawrence	7 + 1 focus group
Estuary (Ste.-Flavie) of the St. Lawrence	5 + 1 focus group
	38 interviews and 5 focus groups

4.2 Discussion

The main goal of the CCC-CURA project was to develop tools and demonstrate that it was possible for communities to plan for resilience through PAR. By examining the process of resilience planning in different communities, it was felt that it would be possible to extract commonalities in terms of steps or processes to consider in developing adaptation plans in communities. As communities were not homogeneous, it was expected that only a few elements might be common. Communities are more than a sum of individuals, and interactions among them can be of a different nature. The capacity to assess the connections among people and groups within a territory can help better coordinate actions. Linking organizations together within a territory may seem straightforward, but can represent a huge challenge, especially for smaller organizations or communities, where expertise and skills are not always present. Social network analysis is a convenient tool that can help organizations to better understand the connections among groups within a territory and how their interactions can be used for a common goal (Crona and Parker 2012). In this project, a guide was developed with and for NGOs. In addition, some of the tools were presented in the summer (Notre-Dame-du-Portage, QC) and winter (Moncton, NB), in both French and English, to make them more accessible to various groups.

As previously mentioned, there are many guides and tools for assessing vulnerabilities, defining adaptation strategies, and developing resilience or adaptation plans. Finding some that are accessible to small communities was an important ambition in this project and one of the reasons for developing the resilience planning kit (Vasseur 2012). Despite select communities that had some similar characteristics, it was obvious by examining the profiles that they all differed. In addition, the relationships among people and groups in each community might have been a factor for not also selecting the same approach to develop adaptation strategies.

Adaptation strategies should be developed at the local level because of the challenges related to heterogeneity and complexities of communities. While some

4.2 Discussion

governments prefer a top-down approach and tend to be outcome-oriented, with no concern related to social acceptance, others prefer a bottom-up approach. As Sherman and Ford (2013, p. 419) have suggested: "In bottom-up approaches, community-based institutions and local people carry out the design and implementation of a project, often with empowerment and capacity-building as key objectives." So, in CCC-CURA, PAR contributed to a more bottom-up approach, although local governments were in many cases involved.

Planning such activities in so many communities can be quite challenging and has certainly reduced the capacity to duplicate some of the good practices. There are many lessons learned from this project and they are discussed later, in Chap. 7. A few can, however, be mentioned here. For example, reaching a good representation of a community can be difficult. In many cases, people tend to refer to friends or people having the same values or attitudes toward a subject. Social representations of an issue like climate change can cause groups to form based on common interests (Breakwell 1993). It is important to also note that some participants came with their personal agenda and were not completely open to some other community members' issues. Planning meetings was also not easy, with most people in small communities being involved in several activities. Finally, mobilization tended to decline over time, since no new storms occurred during the course of the interventions.

References

Adger WN (2003) Social capital, collective action, and adaptation to climate change. Econ Geogr 79(4):387–404

Beaulieu N, Jaramillo J, Leclerc G (2002) The vision-action-requests approach across administrative levels: a methodological proposal for the strategic planning of rural development. Internal report, CIAT/MTD, Cali/ Montpellier, 30 pp

Beaulieu N, Santos Silva J, Plante S (2016) Using a vision of a desired future in climate change adaptation planning: lessons learned in the municipality of Rivière-au-Tonnerre (Québec, Canada). Clim Dev 8(5):447–457

Bennett NJ, Blythe J, Tyler S, Ban NC (2016) Communities and change in the anthropocene: understanding social-ecological vulnerability and planning adaptations to multiple interacting exposures. Reg Environ Change 16(4):907–926

Breakwell GM (1993) Social representations and social identity. Papers Soc Represent 2:1–20

Cargo M, Mercer SL (2008) The value and challenges of participatory research: strengthening its practice. Annu Rev Public Health 29:325–50

Carpenter SR, Walker BH, Anderies JM, Abel N (2001) From metaphor to measurement: resilience of what to what? Ecosystems 4:765–781

Crona BI, Parker JN (2012) Learning in support of governance: theories, methods, and a framework to assess how bridging organizations contribute to adaptive resource governance. Ecol Soc 17(1):32

Fals-Borda O (1987) The application of participatory action research in Latin America. Int Sociol 2:329–347

Fazey JA, Fischer J, Sherren K, Warren J, Noss RF, Dovers SR (2007) Adaptive capacity and learning to learn as leverage for social-ecological resilience. Frontiers Ecol Environ 5:375–380

Mangoyana RB, Thomsen DC, Smith TF, Preston BL, Heinz S, Maloney M, Withycombe G Armstrong I (2012) Prioritising coastal adaptation development options for local government. Literature review of adaptation to climate change in the coastal zone. Coastal Adaptation Decision Pathways Project, Australian Department of Climate Change and Energy Efficiency, 35 pp

Mapfumo P, Adjei-Nsiah S, Mtambanengwe F, Chikowo R, Giller KE (2013) Participatory action research (PAR) as an entry point for supporting climate change adaptation by smallholder farmers in Africa. Environ Dev 5:6–22

McSweeney K, Coomes OT (2011) Climate-related disaster opens a window of opportunity for rural poor in northeastern Honduras. PNAS 108:5203–5208

Reininger BM, Rahbar MH, Lee M, Chen Z, Alam SR, Pope J, Adams B (2013) Social capital and disaster preparedness among low income Mexican Americans in a disaster prone area. Soc Sci Med 83:50–60

Schechtman J, Brady M (2013) Cost-efficient climate change adaptation in the North Atlantic (Final Report). National Oceanic and Atmospheric Administration, Sea Grant and North Atlantic Regional Team, 253 pp

Sherman MH, Ford J (2013) Stakeholder engagement in adaptation interventions: an evaluation of projects in developing nations. Clim Pol 14:417–441

Smith MS (2016) Advances in Guidance Standards for Adaptation Planning. Science and Solutions for Australia and the Global Community. CSIRO. 10 pp

Statistics Canada (2011) 2011 Census of population. http://www12.statcan.gc.ca/census-recensement/index-eng.cfm. Accessed 21 Aug 2014

Vasseur L (2011) Moving from research into action on issues of climate change for a Canadian community: integration of sciences into decision making. Int J Clim Change Impacts Responses 2:115–126

Vasseur L (2012) Getting started with community resilience planning. *A Kit to Implement Dialogue on Planning Community Resilience to Environmental and Climate Changes*. Training manual. 21 pages (23 pages in French). Produced for Coastal Communities Challenges (SSHRC-CURA), Southern Gulf of St. Lawrence Coalition for Sustainability and the Regional Adaptation Collaboratives (RAC) Climate Change Program

Vasseur L, Plante S, Znajda SK, Thornbush M (submitted) How coastal community members perceive resilience: a case from Canada's Atlantic Coast. Ambio

Yin RK (2003) Cases study research: design and methods. Applied Social Research Methods Series 5, Sage Publications, 282 pp

Chapter 5
Findings from Initial Interviews

Abstract The initial set of interviews took place soon after the winter storms, in December 2010 and January 2011. The elaboration of the scheme of interview was collectively executed with Coastal Community Challenges-Community-University Research Alliance (CCC-CURA) partners (researchers, municipalities, citizens, nongovernmental organizations or NGOs, etc.) and pretested in the Acadian Peninsula, with the participation of the Coastal Zones Research Institute Inc. (CZRI) and the Université de Moncton, Shippagan Campus (New Brunswick). Residents of rural coastal communities were interviewed in their native tongue (French or English). Based on semi-directed interviews held both singly and in couples, data analysis revealed several findings from the initial interviews that are relayed in this chapter. In particular, those affected by the 2010 winter storms in the Atlantic provinces (Québec, New Brunswick, and Prince Edward Island) of Canada reported experiences and impacts that were mainly associated with storm surge and flooding as well as high waves and coastal erosion. The impacts affected personal property as well as businesses and public infrastructure, including roads, graveyards, and bridges, influencing evacuation points and the emergency response. Even though most people lived at the coast most of their lives and this was not the only major storm that they had experienced, their responses mainly included emotional reactions (of fear, stress or worry, panic, powerlessness), especially for those who were cut off from the mainland. The elderlies were particularly fearful of having to leave their homes and/or communities and this is indicative of the importance of demographics affecting experiences and responses. Various stressors were expressed by people and panic set in when people realized the extreme damages. In some cases, participants expressed powerlessness. There was a tendency to forget past storms, potentially as a psychological coping mechanism.

Keywords Experiences · Responses · Major storms · Psychosocial barriers

5.1 Experience with 2010–2011 Major Storm Events

Thirty-four respondents noted in the interviews that they had been affected by the 2010 storms, and the impacts were felt either at the community or personal level. In the Morell district of Prince Edward Island (PEI), two respondents mentioned the loss of equipment from storm surge, high water levels entering structures, erosion of a causeway, and blueberry fields inundated by saltwater. In the Northumberland Strait region of New Brunswick (NB), eight respondents described damage to infrastructure, such as roads and marina docks, due to wave surge and high water levels. Six respondents in Ste.-Flavie, Québec (QC) described flooding, erosion, blocked roads, damage to infrastructure (such as retaining walls and septic systems), and the evacuation of the community by 30 families.

The communities of Ste.-Marie-St.-Raphael, NB and Maria and Bonaventure, QC, in particular, were strongly affected by the storms, with 18 respondents describing tides rising to the level of their houses and moving large pieces of debris (such as rocks and tree limbs); erosion of up to 15 m of their property; flooding in the basements of primary and/or secondary homes (chalets or cottages); damage to breakwaters and retaining walls; and evacuation of residents in high-risk areas due to high-water levels. One respondent in Baie des Chaleurs, QC decided to move his house after the 2010 storms to a planned, prepared, and available site for those interested in moving. Those interviewed in Ste.-Flavie, QC, one of the hardest hit communities, described the psychological distress that many community members experienced during and after this set of storms and in anticipating similar future major events.

5.2 Psychosocial Barriers to Change

Out of 44 of 74 respondents who experienced the 2010 storms, five major emotional responses were evident, including fear, stress or worry, panic, powerlessness, and lack of stress. Fear was provoked knowing that residents were "cut off from the mainland" (e.g. Cocagne), where there was access to provisions (where food could be purchased), healthcare, and services (e.g., lottery tickets). People were also fearful having examined maps denoting areas of flooding in scenarios of climate change (in PEI and Shippagan) as well as damage from larger storms, particularly from rocks and debris (Pigeon Hill, Acadian Peninsula). There were also fears voiced about whether there would be access to potable water (Northumberland Strait) and of rising waves without being able to retreat away from the shoreline due to a lack of emergent land (Ste.-Marie-St.-Rachaël), *"the sea breaks on my land and I don't know what to do, I'm scared, the windows will break, rocks are being thrown to the windows"* (translated by L Vasseur; Acadian, Pigeon Hill, affected by 2010). Elderly respondents were especially fearful of having to leave their home or community; and there was another age-related fear expressed by young people

5.2 Psychosocial Barriers to Change

(e.g., 18 year-olds) about not staying in Shippagan. In Maria, for instance, many people already relocated because they could not withstand the pressure of living there, "*in fact, we finished by understanding that if they didn't have the nerves to stay there, the only solution was to move. This means that we have lost a few people who decided to move, to be in a location... far from water bodies*" (translated by L Vasseur; Baie des Chaleurs, affected by 2010).

Stress or worry were expressed in terms of how to pay for damage and the presence of young children (Northumerland Strait); increased frequency of storms (Cocagne); an inability to sleep at night in anticipation and building anxiety of hearing waves crash against the breakwater and hearing pumps working all night in the basement (Baie des Chaleurs); or, alternately, of power going out and losing equipment. Civil servants (municipal workers) were stressed because people sought them out for support and answers (e.g., Maria): "*It's to live with this. It's a huge challenge. Live with this and control my stress. It's because each time we have this situation, all eyes are on me. It's my concern*" (translated by L Vasseur; Baie des Chaleurs, not personally affected by 2010).

Panic ensued when damages were extreme. People also felt powerless in the face of storms and incumbent damages, such as those not affected by the 2010 storms, but affected by past storms in terms of flooding in PEI:

I feel pretty powerless about changing any of it, because I'm not a farmer, so I don't have a voice in that respect. I'm not a dyed in the wool environmentalist, so I don't have a voice in that respect. I'm just a guy out there trying to stay alive, trying to keep his head above water in terms of your day to day things. And wondering if it's making any kind of a difference or not.

The scale of things that could happen is so beyond human comprehension. It's one thing to say and one thing to see in a movie and one thing to see in a computer simulation, the hillside sliding into the river, it's another thing for it to actually happen. I think the chances of it happening are pretty slim so I don't let it keep me up at night. But when the topic comes around, yeah, I'm just as worried as the next guy. Maybe a little more because I try to do a little more to have a little less impact (translated by SK Znajda).

Some respondents that were previously affected by storms also expressed a feeling of powerlessness. Interestingly, a few people were not stressed and/or did not perceive others in their community to be stressed because they were accustomed to experiencing waves and wind, etc. associated with storms (as in Ste.-Marie-St.-Raphaël and Shippagan), but these individuals either had not been previously affected or could not recall impacts from previous storms. This could be construed as an adaptive psychosocial mechanism used involuntarily in order to cope with traumatic events of the past (event memory). The downfall to this natural (psychological) approach is that it counteracts any action that is necessary for real-world adaptation to the problem.

5.3 Discussion

Zander et al. (2013) have examined the option to relocate for aboriginal people located in coastal communities in northern Australia (same for Vasseur and Tremblay 2014 for the Elsipogtog community situated in the Acadian Peninsula of Canada). Their study respondents had all heard about climate change and some (48%) had already seen associated environmental change (e.g., sea-level rise). For safety reasons, most respondents chose relocation as an adaptation strategy; however, not all agreed:

(1) Of respondents who would consider relocation (58%), most preferred to remain close to the sea and many considered moving inland so long as community facilities would also be relocated; and
(2) Others saw it as unlikely that they would relocate and would prefer to adapt where they were, with the provision of government supports (including shelters, defenses, and infrastructure).

Because relocation is not an option that suits everyone, such as the elderly, better preparatory (monitoring, evacuation, and emergency) and planning (longer term, including hazard-proofing and governance) measures need to be in place in order to improve adaptation of rural coastal communities to major storms occurring in Atlantic Canada. The monitoring of storms, for instance, could improve evacuation and (immediate) emergency responses. Models are already being used, as for instance to track autumn upwelling events in the Alaskan Beaufort Sea (Pickart et al. 2011) that lead to a better understanding of trigger points and the local-to-regional response. Such models have already been applied to examine the effects of tides and storm surge on the coastal water energy budget in the Northwest Atlantic (e.g., Bernier and Thompson 2010). Satellite observations are instrumental to inform these models as well as relay data concerning sea-surface temperature, for example, affecting storm-induced changes (e.g., Han et al. 2012).

Long-term planning responses for hazard-proofing, for instance, need to be effective in order to resolve local-to-regional issues, as evoking a "building with nature" response along coasts in The Netherlands (van Slobbe et al. 2013). This type of coastal protection strategy can be effective for dealing with the increasing risks associated with storm surge-induced floods as well as sea-level rise and land subsidence. Working with nature is a more responsive (robust) and flexible (adaptive and sustainable) approach that is yet cost-effective. It was found in this study that although people prefer to build some hard defenses (e.g., seawalls, Snoussi et al. 2009) as "typical measures" (Jonkman et al. 2013), in Atlantic Canada this is often not feasible due to a lack of government funding or land-use regulations.

Quantitative approaches to study spatial-temporal dynamics affected by climate change have been espoused (e.g., Magris et al. 2014 for marine conservation planning) and further research of this type is needed. Integrated approaches are preferred, as with integrative deliberative procedures, that involve actors like policymakers as well as scientists and the general public (Kane et al. 2014). Integration

5.3 Discussion

needs to also occur at the disciplinary level, as advocated by le Cornu et al. (2014) when they define coastal and ocean planning from a broad framework encompassing biophysical and social attributes. By linking resilience and social well-being (cf. Armitage et al. 2012), it is possible to foster a more realistic (complex) social-ecological perspective for policy making and management that is informed by human values and agency as well as part of interdisciplinary research (Mustelin et al. 2013) and governance. Such an interdisciplinary and integrated approach has been extended to coastal planning (e.g., Lloyd et al. 2013). Longitudinal studies, however, are still needed in order to verify poorer health outcomes and the stressors triggering these, such as possibly the capacity for helping others to adapt to change (Hogan et al. 2013). Helping others, however, does not always transmit to coordinated preventative strategies and links to local government could also be lacking (Linnekamp et al. 2011).

When a central authority is missing or inactive in governance, problems may result that stem from growth-related issues and management (as with the lack of plans) and complex multijurisdictional regulatory frameworks may help support sectoral development, as with Arctic shipping (Dawson et al. 2014). Working in partner-oriented research or research partnerships that (from the outset) stimulate educational outreach and integrated community action contribute toward collaborative approaches (Adams et al. 2014; Frazier et al. 2010) that are already evident among indigenous groups and enable for a relevant social and cultural context (Housty et al. 2014). In this case, co-governance is possible so that traditional knowledge and experiences may affect planning and management (Jones et al. 2010). When indigenous people are able to detect environmental change, it impacts their resource use and management (Aswani and Lauer 2014).

For adaptation planning, in particular, it is important to identify (and map) climatic vulnerabilities in order to assist in the prioritization that guides the process (Okey et al. 2014). Part of this marine spatial planning requires communication and engagement of communities (Halpern et al. 2012), including gauging the perceptions of QC residents that are rarely considered in coastal risk and adaptation analysis in the Gulf of St. Lawrence (Friesinger and Bernatchez 2010). According to Sahin et al. (2013), a multistakeholder approach for prioritization conveys a difference in opinion regarding improved building design (preferred by residents and politicians) and raising public awareness (preferred by experts). Gauging local people's perceptions as well as any relevant statistics (i.e., demographics) allows for the identification of issues and targets as well as the determination of any local changes and trends that have been (or should be) monitored (Vlasova and Volkov 2013). These perceptions might be affected by high-impact events, such as storms and flooding, which are site- and time-specific (Muir et al. 2014).

Actor perception affects the adaptive capacity of communities, with generic determinants being awareness and funding and more specific determinants tending to be pragmatic and context-based, including sector-specific determinants (Richards et al. 2013; Sovacool 2012). Affecting perceptions are also low levels of education, as in Pakistan, where the adaptive capacity of local populations is negatively affected by this (Salik et al. 2015). In addition, the response strategies employed at the household-to-community level as well as any economic opportunities created are vital to community resilience from a social-ecological system (SES) viewpoint, as seen in Sami fishing communities (Broderstad and Eythórsson 2014). The transmission of local and indigenous knowledge could bolster community resilience, as for coastal and small island communities (Hiwasaki et al. 2015). Fostering positive social relationships, as with communities, and involving multiple actors has been found to be an effective management approach that is part of co-management (Marín et al. 2012).

It is important to involve bottom-up approaches to governance that consider the inputs and participation of community members. One of the reasons to support this claim is made by Hopkins et al. (2011), who argue that information is rendered more useful to end-users when they are involved somehow through participatory communication. Frameworks can be either analysis-oriented or action-oriented (Binder et al. 2013), and the latter is potentially more relevant for a PAR approach that tends to motivate action for change. Social learning is possible given such an approach, with learning experiences possible via attendance in participatory workshops (Johnson et al. 2012; Rodela 2011). Such a participatory approach empowers community members with skills that can build on community knowledge, innovation, and resilience (Smith et al. 2011). According to Carmack et al. (2012) "resilience thinking" can develop with the participation of small coastal communities in the observation, adaptation, and transformation of their SESs.

There are also opportunities for social-ecological change that allow for transformation through flexible (and innovative, cf. Olsson et al. 2006) governance (Folke et al. 2011; Olsson et al. 2004). This type of approach can be construed as part of "futures research" that aims to reduce vulnerability and augment adaptive capacity, and has already been applied to deal with environmental issues and problems (Bengston et al. 2012), also possibly from a participatory approach (Gidley et al. 2009). Traditionally, as part of sustainability research, "adaptive inference" has adopted a long-term perspective, as over the course of 30 years (Holling and Allen 2002), to examine ecosystem resilience. Suck work by Holling (1992, 1994; Walters and Holling 1990) originally developed notions of ecosystem dynamics, resilience, and recovery that are vital for sustainability management.

The most difficult aspect of a changing climate is uncertainty. This is expressed and developed in various Canadian research studies, as for example by Grima (1993) for the Great Lakes region in terms of water resource management. Petersen et al. (2013) propose a proactive approach to climate change adaptation in the Great Lakes region that is lacking in most of the individuals that participated in their study. Nevertheless, the Great Lakes region has been exemplary for successful adaptive management (Gronewold et al. 2013). Elsewhere in Canada, models have provided a basis for simulations of landfalling autumn storms affected by cyclone activity that has been shifting poleward with slightly decreased intensities (Perrie et al. 2010; also Lozano et al. 2004 for the European Atlantic coast). In Newfoundland, Hurricane Florence in September 2006 affected the Atlantic coast continental shelf by propagating a sea-level disturbance with high-amplitude oscillations (Thiebaut and Vennell 2010). Simulation models have conveyed, for northern British Columbia, Canada (Pitcher and Ainsworth 2010), and dexterity (of adapting fishing gear) as particularly beneficial to cope with the increased risks associated with climatic variability. The Bonne Bay region located on the west coast of Newfoundland (Lowitt 2014) has provided an example of changing fisheries directed at improved resilience and enhanced (democratic) food security. Biophysical vulnerability was assessed in the Bay of Fundy (Tibbetts and van Proosdij 2013) to changing tidal levels and storm surge using a geomatics tool (in ArcGIS 9.3) prior to any events, so that coastal managers and planners could implement measures to reduce vulnerability and enhance the adaptive capacity of local communities.

5.4 Experiences and Lessons Learned from Coastal Storms

People who are accustomed to living at the coast are willing to accept both the positive and negative aspects of their choice. Ueda and Torigoe (2012) show that victims of tsunamis tend to accept their vulnerability as a condition of accessing the fertility of the sea. Authors have noted the ongoing conflict of occupation versus risks at the coast, such as at the Praia de Faro in the Portuguese southern coast (Costas et al. 2015). These authors report that risk perception in Portugal is affected by place attachment as well as underestimated impacts/lessons learned and cultural aspects in addition to the imposition of relocation measures as part of risk mitigation. Behavioral change is key to adapting at various levels and is seen as reflecting personal and/or cultural attachments that may be occupation-based, as for example of fishers in resource-dependent communities (Forster et al. 2014).

There is a recent emerging strand of research that recognizes socially constructed knowledge as part of social representation that is based on understandings by subgroups (e.g., Moloney et al. 2014); and this could affect the experience of different cultural groups as well as age-based and gender-specific groups within communities. For instance, risks may be socially constructed so that risk communication needs to be integrated into the entire risk governance process, from pre-assessment to appraisal to characterization/evaluation and finally to management (Kane et al. 2014).

The perception and responses of locals provide important (site-specific) inputs to sustainable (adaptive) management, allowing for the identification of needs as well as challenges and opportunities that can support effective decision making (cf. Munji et al. 2014). For instance, a series of workshops with residents (in Boston, Massachusetts, Douglas et al. 2012) reveals a lack of knowledge about resources and adaptation because community members are not included in the planning process, even though there is a general sentiment of eagerness to learn and become actively involved in the decision-making process. Among the lessons learned in their work was that local residents should be engaged early in the planning process, as this can foster trust and consensus (enabling progression from the conceptual to implementation stages) as well as offer opportunities for education. It is known (e.g., MacInnis et al. 2015) that American public opinion, for instance, can be influenced by endorsement, as when preparation is endorsed by researchers or government (rather than religious or business leaders), and this may be an issue of trusting people in authority. The availability of information (to residents and decision makers) is another factor that affects perception (Dempsey and Fisher 2005).

Individuals were mostly self-sufficient in their experiences of the 2010 storms, although there was some feeling of community cohesion. Others, however, felt that they could have drawn on others (friends and neighbors) for help. However, having experienced storms previously better prepared individuals so that they became more attentive to changing weather conditions and had an action plan for their responses at an individual level. This was not evident at the community level, as individuals identified a lack of administrative organization and support, especially in the form of emergency plans, and there is much work that can be executed in community-regional planning for storms. Such a lack of planning can be considered maladaptive. Macintosh (2013) observes that when coastal urban planners do not attend to lower level responsibilities, and governmental frameworks are not clear, this can lead to ambiguous decision guidelines. This is especially tricky during disasters, when plans need to be in effect and well-communicated so that they can be executed at various levels, including emergency evacuation plans at the community level.

There is the problem of poor communication infrastructure, as some people did not have telephone and could not contact others for help. In case of an emergency, when it is not enough to have a contact number for help, people actually need to be able to access telephones, radio, and other forms of communications media. The following, for instance, was suggested by a man from QC: *"Wants an amateur radio club in Riviere du Grain—for improved communication. E.g. can call ambulances etc. People in area don't have telephones. One has a cell phone with an antennae"* (translated by SK Znajda).

One of the first disaster risk reduction (DRR) actions suggested by people was effective and proactive communication. Weather forecasting and monitoring appear to be improving in Atlantic Canada, and some participants noted weather observations for various stations used for local-regional weather forecasting: *"People pay more attention to the weather now. Before getting on the ferry to Gaspe they check not just one weather station but many. People think twice before building big houses by the sea, about distance away from water. People don't want to buy close to the sea anymore, because of the risks"* (translated by SK Znajda; 50+ year-old woman from NB, affected by 2010 storms). So, not only are people now more alert to weather forecasts, there is also improved communication about weather: *"Information about weather is well circulated. For past 5–6 years check weather network every morning before heading out"* (translated by SK Znajda; 57 year-old man from NB).

The emergency response was particularly limited in these communities and needed to be better orchestrated (e.g., risk funds reserve). Local knowledge is increasingly recognized as advantageous to inform preparation and the planning process. It has been deployed in order to enhance adaptive capacity in various ways, including for building protection structures, re-vegetation works, self-sufficiency practices, and transferring knowledge across generations (McNamara and Westoby 2011). Aswani and Lauer (2014) and others (e.g., the Haida Nation as a partner in marine conservation, Jones et al. 2010), for instance, have highlighted the role of local ecological knowledge acquired from indigenous people. This approach of integrating local knowledge has been recognized as a resource benefiting the adaptive capacity of communities to climate change, as through knowledge co-production (Boillat and Berkes 2013). When indigenous knowledge is not integrated, there is the possibility of reduced resilience and, therefore, vulnerability (cf. Kalanda-Joshua et al. 2011). Participatory research methods ensure that knowledge sharing can occur (as when links are established with scientific and indigenous communities) to divulge coping strategies and early warning systems, for instance, and enhance the adaptive capacity and build resilience (Valdivia et al. 2010).

Communities also need to be more connected. Rather than considering the individual and community in isolation, many respondents felt that community cooperation was needed, as with higher level local-to-regional planning for the emergency response. Indeed, community vitality has been espoused to be a cornerstone in sustainable development (Dale et al. 2010). Other research has noted the need for better communications between community and government agencies (Myers et al. 2012) that improves inter-connectivity as well as perhaps even trust. Households that have a greater "social connectivity," which is affected by gender, age of household head, and household size, enhance their resilience (Cassidy and Barnes 2012). Education is also instrumental in this regard, as information and knowledge transfer between actors come from different sectors (e.g., elected, public and economic sectors, civil society, and NGOs) may facilitate multilevel governance leadership and action (e.g., stewardship). An important element of effective adaptation planning has been the interplay between bottom-up and top-down approaches (multilevel governance), with the former evoking participation in adaptation design, implementation, and monitoring and the latter operative for strategy prioritization and legitimization (Bizikova et al. 2014). There is also a recognition of the need to improve resilience capacity at the municipal level in order to develop sub-national adaptation protocols, as for instance is evident in Maine (Camill et al. 2012), that work for communities at the local-to-regional scale.

5.5 Lessons Learned and Additional Measures

Suggested future measures addressed the importance of research and education (including reducing environmental impact by installing solar panels, composting, gardening, recycling); building an emergency response plan; preparing in advance for storms; not being alone (knowing that other areas are also affected); and realizing the value of teamwork. Public servants, in particular, had a very specific perspective in terms of lessons learned and an action plan. Some of their views included:

- Improving information flow—as for instance better documenting and the use of materials (photographs) to show damage and educate people; improved communications among local service districts (LSDs); conveying complex information to the public (as through a communications specialist), such as maps, to raise awareness at the community level.
- Pursuing incorporation status—for the development of better guidelines (as in Cocagne).
- Practicing the principles of sustainable development (Acadian Peninsula).
- In Maria specifically, double-checking permits and directly explaining risks; rebuilding infrastructure differently to withstand future storms; flood-risk mapping; spending less time convincing people to evacuate.
- Finally, in Ste.-Flavie, their experience with governmental decree left sociopsychological effects in the community and exodus of families.

At the community level, it is important to build houses further back from the coast and rethink about building near water; taking better precautions; and realizing whether protection may be needed, such as rock/retention walls. Government-level measures are also needed (higher up than the individual or community level) for various actions, such as: engaging LSD decision-makers in climate change discussions (Cocagne); starting recycling services (Cap Bateau, QC); or effective management of vacant sites and relocation of properties and infrastructure (Ste.-Flavie, QC); plans; more training at the municipal level; and the enactment of emergency-training exercises.

References

Adams MS, Carpenter J, Housty JA, Neasloss D, Paquet PC, Service C, Walkus J, Darimont CT (2014) Toward increased engagement between academic and indigenous community partners in ecological research. Ecol Soc 19(3):5

Armitage D, Béné C, Charles AT, Johnson D, Allison EH (2012) The interplay of well-being and resilience in applying a social-ecological perspective. Ecol Soc 17(4):15

Aswani S, Lauer M (2014) Indigenous people's detection of rapid ecological change. Conserv Biol 28(3):820–828

Bengston DN, Kubik GH, Bishop PC (2012) Strengthening environmental foresight: potential contributions of futures research. Ecol Soc 17(2):10

Bernier NB, Thompson KR (2010) Tide and surge energy budgets for Eastern Canadian and Northeast US waters. Cont Shelf Res 30:353–364

Binder CR, Hinkel J, Bots PWG, Pahl-Wostl C (2013) Comparison of frameworks for analyzing social-ecological systems. Ecol Soc 18(4):26

Bizikova L, Crawford E, Nijnik M, Swart R (2014) Climate change adaptation planning in agriculture: processes, experiences and lessons learned from early adapters. Mitig Adapt Strateg Glob Change 19:411–430

References

Boillat S, Berkes F (2013) Perception and interpretation of climate change among Quechua farmers of Bolivia: indigenous knowledge as a resource for adaptive capacity. Ecol Soc 18(4):21

Broderstad EG, Eythórsson E (2014) Resilient communities? Collapse and recovery of a social-ecological system in Arctic Norway. Ecol Soc 19(3):1

Camill P, Hearn M, Bahm K, Johnson E (2012) Using a boundary organization approach to develop a sea level rise and storm surge impact analysis framework for coastal communities in Maine. J Environ Stud Sci 2:111–130

Carmack E, McLaughlin F, Whiteman G, Homer-Dixon T (2012) Detecting and coping with disruptive shocks in Arctic marine systems: a resilience approach to place and people. Ambio 41:56–65

Cassidy L, Barnes GD (2012) Understanding household connectivity and resilience in marginal rural communities through social network analysis in the village of Habu, Botswana. Ecol Soc 17(4):11

Costas S, Ferreira O, Martinez G (2015) Why do we decide to live with risk at the coast? Ocean Coast Manage 118(Part A):1–11

Dale A, Ling C, Newman L (2010) Community vitality: the role of community-level resilience adaptation and innovation in sustainable development. Sustainability 2:215–231

Dawson J, Johnston ME, Stewart EJ (2014) Governance of Arctic expedition cruise ships in a time of rapid environmental and economic change. Ocean Coast Manage 89:88–99

Dempsey R, Fisher A (2005) Consortium for Atlantic regional assessment: information tools for community adaptation to changes in climate or land use. Risk Anal 25(6):1495–1509

Douglas EM, Kirshen PH, Paolisso M, Watson C, Wiggin J, Enrici A, Ruth M (2012) Coastal flooding, climate change and environmental justice: identifying obstacles and incentives for adaptation in two metropolitan Boston Massachusetts communities. Mitig Adapt Strateg Glob Change 17:537–562

Folke C, Jansson Å, Rockström J, Olsson P, Carpenter SR, Chapin FS III, Crépin A-S, Daily G, Danell K, Ebbesson J, Elmqvist T, Galaz V, Moberg F, Nilsson M, Österblom H, Ostrom E, Persson Å, Peterson G, Polasky S, Steffen W, Walker B, Westley F (2011) Reconnecting to the biosphere. Ambio 40:719–738

Forster J, Lake IR, Watkinson AR, Gill JA (2014) Marine dependent livelihoods and resilience to environmental change: a case study of Anguilla. Mar Policy 45:204–212

Frazier TG, Wood N, Yarnal B (2010) Stakeholder perspectives on land-use strategies for adapting to climate-change-enhanced coastal hazards: Sarasota, Florida. Appl Geogr 30:506–517

Friesinger S, Bernatchez P (2010) Perceptions of Gulf of St. Lawrence coastal communities confronting environmental change: hazards and adaptation, Québec, Canada. Ocean Coast Manage 53:669–678

Gidley JM, Fien J, Smith J-A, Thomsen DC, Smith TF (2009) Participatory futures methods: towards adaptability and resilience in climate-vulnerable communities. Environ Policy Gov 19:427–440

Grima APL (1993) Enhancing resilience in Great Lakes water levels management. Int J Environ Stud 44:97–111

Gronewold AD, Fortin V, Lofgren B, Clites A, Stow CA, Quinn F (2013) Coasts, water levels, and climate change: a Great Lakes perspective. Clim Change 120:697–711

Halpern BS, Diamond J, Gaines S, Gelcich S, Gleason M, Jennings S, Lester S, Mace A, McCook L, McLeod K, Napoli N, Rawson K, Rice J, Rosenberg A, Ruckelshaus M, Saier B, Sandifer P, Scholz A, Zivian A (2012) Near-term priorities for the science, policy and practice of Coastal and Marine Spatial Planning (CMSP). Mar Policy 36:198–205

Han G, Ma Z, Chen N (2012) Hurricane Igor impacts on the stratification and phytoplankton bloom over the Grand Banks. J Marine Syst 100–101:19–25

Hiwasaki L, Luna E, Syamsidik Marçal JA (2015) Local and indigenous knowledge on climate-related hazards of coastal and small island communities in Southeast Asia. Clim Change 128:35–56

Hogan A, Tanton R, Lockie S, May S (2013) Focusing resource allocation-wellbeing as a tool for prioritizing interventions for communities at risk. Int J Environ Res Public Health 10:3435–3452

Holling CS (1992) Cross-scale morphology, geometry, and dynamics of ecosystems. Ecol Monogr 62(4):447–502

Holling CS (1994) Simplifying the complex: the paradigms of ecological function and structure. Futures 26(6):598–609

Holling CS, Allen CR (2002) Adaptive inference for distinguishing credible from incredible patterns in nature. Ecosystems 5:319–328

Hopkins TS, Bailly D, Støttrup JG (2011) A systems approach framework for coastal zones. Ecol Soc 16(4):25

Housty WG, Noson A, Scoville GW, Boulanger J, Jeo RM, Darimont CT, Filardi CE (2014) Grizzly bear monitoring by the Heiltsuk people as a crucible for First Nation conservation practice. Ecol Soc 19(2):70

Johnson KA, Dana G, Jordan NR, Draeger KJ, Kapuscinski A, Olabisi LKS, Reich PB (2012) Using participatory scenarios to stimulate social learning for collaborative sustainable development. Ecol Soc 17(2):9

Jones R, Rigg C, Lee L (2010) Haida marine planning: First Nations as a partner in marine conservation. Ecol Soc 15(1):12

Jonkman SN, Hillen MM, Nicholls RJ, Kanning W, van Ledden M (2013) Costs of adapting coastal defences to sea-level rise-new estimates and their implications. J Coast Res 29(5):1212–1226

Kalanda-Joshua M, Ngongondo C, Chipeta L, Mpembeka F (2011) Integrating indigenous knowledge with conventional science: enhancing localised climate and weather forecasts in Nessa, Mulanje, Malawi. Phys Chem Earth 36:996–1003

Kane IO, Vanderlinden J-P, Baztan J, Touili N, Claus S (2014) Communicating risk through a DSS: a coastal risks centred empirical analysis. Coast Eng 87:240–248

Le Cornu E, Kittinger JN, Koehn JZ, Finkbeiner EM, Crowder LB (2014) Current practice and future prospects for social data in coastal and ocean planning. Conserv Biol 28(4):902–911

Linnekamp F, Koedam A, Baud ISA (2011) Household vulnerability to climate change: examining perceptions of households to flood risks in Georgetown and Paramaribo. Habitat Int 35:447–456

Lloyd MG, Peel D, Duck RW (2013) Towards a social-ecological resilience framework for coastal planning. Land Use Policy 30:925–933

Lowitt KN (2014) A coastal foodscape: examining the relationship between changing fisheries and community food security on the west coast of Newfoundland. Ecol Soc 19(3):48

Lozano I, Devoy RJN, May W, Andersen U (2004) Storminess and vulnerability along the Atlantic coastlines of Europe: analysis of storm records and of a greenhouse gasses induced climate scenario. Mar Geol 210:205–225

MacInnis B, Krosnick JA, Abeles A, Caldwell MR, Prahler E, Dunne DD (2015) The American public's preference for preparation for the possible effects of global warming: impact of communication strategies. Clim Change 128:17–33

Macintosh A (2013) Coastal climate hazards and urban planning: how planning responses can lead to maladaptation. Mitig Adapt Strateg Glob Change 18:1035–1055

Magris RA, Pressey RL, Weeks R, Ban NC (2014) Integrating connectivity and climate change into marine conservation planning. Biol Conserv 170:207–221

Marín A, Gelcich S, Castilla JC, Berkes F (2012) Exploring social capital in Chile's coastal benthic comanagement system using a network approach. Ecol Soc 17(1):13

McNamara KE, Westoby R (2011) Local knowledge and climate change adaptation on Erub Island. Torres Strait. Local Environ 16(9):887–901

Moloney G, Leviston Z, Lynam T, Price J, Stone-Jovicich S, Blair D (2014) Using social representation theory to make sense of climate change: what scientists and nonscientists in Australia think. Ecol Soc 19(3):19–27

References

Muir D, Cooper AG, Pétursdóttir G (2014) Challenges and opportunities in climate change adaptation for communities in Europe's northern periphery. Ocean Coast Manage 94:1–8

Munji CA, Bele MY, Idinoba ME, Sonwa DJ (2014) Floods and mangrove forests, friends or foes? Perceptions of relationships and risks in Cameroon coastal mangroves. Estuar Coast Shelf S 140:67–75

Mustelin J, Kuruppu N, Kramer AM, Daron J, de Bruin K, Noriega AG (2013) Climate adaptation research for the next generation. Clim Dev 5(3):189–193

Myers SA, Blackmore MJ, Smith TF, Carter RW (2012) Climate change and stewardship: strategies to build community resilience in the Capricorn Coast. Australas J Environ 19(3):164–181

Okey TA, Alidina HM, Lo V, Jessen S (2014) Effects of climate change on Canada's Pacific marine ecosystems: a summary of scientific knowledge. Rev Fish Biol Fisheries 24:519–559

Olsson P, Folke C, Berkes F (2004) Adaptive comanagement for building resilience in social-ecological systems. Environ Manage 34(1):75–90

Olsson P, Gunderson LH, Carpenter SR, Ryan P, Lebel L, Folke L, Folke C, Holling CS (2006) Shooting the rapids: navigating transitions to adaptive governance of social-ecological systems. Ecol Soc 11(1):18

Perrie W, Yao Y, Zhang W (2010) On the impacts of climate change and the upper ocean of midlatitude northwest Atlantic landfalling cyclones. J Geophys Res 115(D23110):1–14

Petersen B, Hall KR, Kahl K, Doran PJ (2013) In their own words: perceptions of climate change adaptation from the Great Lakes region's resource management community. Environ Pract 15:377–392

Pickart RS, Spall MA, Moore GWK, Weingartner TJ, Woodgate RA, Aagaard K, Shimada K (2011) Upwelling in the Alaskan Beaufort Sea: atmospheric forcing and local versus non-local response. Prog Oceanogr 88(1–4):78–100

Pitcher TJ, Ainsworth CH (2010) Resilience to change in two coastal communities: using the maximum dexterity fleet. Mar Policy 34:810–814

Richards R, Sanó M, Roiko A, Carter RW, Bussey M, Matthews J, Smith TF (2013) Bayesian belief modeling of climate change impacts for informing regional adaptation options. Environ Modell Softw 44:113–121

Rodela R (2011) Social learning and natural resource management: the emergence of three research perspectives. Ecol Soc 16(4):30

Sahin O, Mohamed S, Warnken J, Rahman A (2013) Assessment of sea-level rise adaptation options: multiple-criteria decision-making approach involving stakeholders. Struct Surv 31(4):283–300

Salik KM, Jahangir S, Zahdi WuZ, ul Hasson S (2015) Climate change vulnerability and adaptation options for the coastal communities of Pakistan. Ocean Coast Manage 112:61–73

Smith TF, Daffara P, O'Toole K, Matthews J, Thomsen DC, Inayatullah S, Fien J, Graymore M (2011) A method for building community resilience to climate change in emerging coastal cities. Futures 43:673–679

Snoussi M, Ouchani T, Khouakhi A, Niang-Diop I (2009) Impacts of sea-level rise on the Morrocan coastal zone: quantifying coastal erosion and flooding in the Tangier Bay. Geomorphology 107:32–40

Sovacool BK (2012) Perceptions of climate change risks and resilient island planning in the Maldives. Mitig Adapt Strateg Glob Change 17:731–752

Thiebaut S, Vennell R (2010) Observations of a fast continental shelf wave generated by a storm impacting Newfoundland using wavelet and cross-wavelet analyses. J Phys Oceanogr 40:417–428

Tibbetts JR, van Proosdij D (2013) Development of a relative coastal vulnerability index in a macro-tidal environment for climate change adaptation. J Coast Conserv 17:775–797

Ueda K, Torigoe H (2012) Why do victims of the tsunami return to the coast? Int J Jap Sociol 21(1):21–29

Valdivia C, Seth A, Gilles JL, García M, Jiménez E, Cusicanqui J, Navia F, Yucra E (2010) Adapting to climate change in Andean ecosystems: landscapes, capitals, and perceptions shaping rural livelihood strategies and linking knowledge systems. Ann Assoc Am Geogr 100(4):818–834

Van Slobbe E, de Vriend HJ, Aarninkhof S, Lulofs K, de Vries M, Dircke P (2013) Building with Nature: in search of resilient storm surge protection strategies. Nat Hazards 66:1461–1480

Vasseur L, Tremblay E (2014) Coastal ecosystem in Kouchibouguac National Park of Canada: adaptation possibilities for protecting traditional knowledge of local a community. In: Buyck C (ed) Safe havens: protected areas for disaster risk reduction and climate change adaptation. IUCN, Gland, Switzerland, pp 33–40

Vlasova T, Volkov S (2013) Methodology of socially-oriented observations and the possibilities of their implementation in the Arctic resilience assessment. Polar Rec 49(250):248–253

Walters CJ, Holling CS (1990) Large-scale management experiments and learning by doing. Ecology 71(6):2060–2068

Zander KK, Petheram L, Garnett ST (2013) Stay or leave? Potential climate change adaptation strategies among Aboriginal people in coastal communities in northern Australia. Nat Hazards 67:591–609

Chapter 6
Findings from Follow-up Interviews

Abstract The Coastal Community Challenges-Community-University Research Alliance (CCC-CURA) project was a longitudinal project that encompassed a second series of interviews in 2014 in order to determine whether there were changes over time in these communities that were related to awareness and actions to adaptation and resilience. The second interviews also examined the role that the CCC-CURA project played in enhancing resilience and governance in these studied communities. The results suggested that, in general, perceptions and attitudes toward extreme events did not really change over time. While, in general, people were more aware of the risks, this did not necessarily translate into action. In both provinces (Québec and New Brunswick), people believed that governments were ready in case of emergency; however, communication for some remained a challenge. Contrary to the first interviews, most people understood resilience and believed that they were resilient. In terms of knowledge of the CCC-CURA project, few were involved and, therefore, benefited from the interventions. The second series of interviews have demonstrated the importance of sustained interventions in order to enhance resilience capacity in a community. Without continuous efforts, people tend to revert back to old habits and few changes occur.

Keywords Longitudinal study · Perception · Preparedness · Social resilience · Natural environment · Barriers to change

6.1 Introduction

Transformation and adaptation in communities can be a long process that can be accelerated through education, capacity-building, or policy development. Experience with storms may also affect how people and communities decide to adapt or not to climate change extreme events in coastal communities. Understanding how knowledge and experience can affect people's perceptions of the hazard and motivate them to act remains an important aspect of developing better tools and strategies to enhance people and communities' capacity and desire to change.

The social dimension of climate change adaptation can be quite complex and affected by the temporal and spatial realities of communities. Their experience regarding storm surge can motivate some changes and enhance adaptation, but this reaction may vary over time depending on the exposure to new hazards. In the Coastal Community Challenges-Community-University Research Alliance (CCC-CURA) project, we wanted to determine whether the project had an effect on people in the community and how people's perceptions changed over time if they were further exposed or not to storms or other extreme events. As explained in the previous chapters, between 2011–2012 and 2014, CCC-CURA researchers acted as facilitators between the different actors (public and economic sectors and civil society) in order to fill the "adaptation deficit" (Burton and May 2004), using various methods to reinforce adaptive capacity and resilience (MEGF, participative mapping, "kitchen assemblies," Open forum, SWOT, focus on climate change issues, focus groups, and individual interviews). This longitudinal study aimed to better understand whether the interventions executed by the CCC-CURA project could help enhance people or their communities' interests to adopt adaptation measures. To do so, follow-up interviews were completed in the same communities where CCC-CURA activities occurred between 2011–2012 and 2014. The methodology was already explained in Chap. 4.

6.2 Results

The interviews suggested that there was no real difference in the perceptions of people who had been affected by the 2010 storms and those not affected, and this remained true in 2014. This statement might explain the results: *"Time arranges things."* This might have been true in Québec (QC) considering that few large storms occurred since 2010. However, in New Brunswick (NB) this may have stemmed from the results of the ice storm in 2013, which caused power outage to all residents, leveling their perceptions. All participants mentioned that climate was changing. This included hotter weather (especially in summers), more high tides, an earlier spring, and more rain in winters. Winters are not as cold and one person mentioned less snow.

When people were first asked what changed since the 2010 storms, most related to their experiences with it or in NB, with a recent ice storm that led to a power outage of several days. One person in NB suggested that the 2010 storms have helped communication, which has since improved. In addition, the community is now talking about emergency plans. With the most recent ice storm of the winter of 2013, it was explained that emergency services and the Red Cross came faster and the provincial government was more present than in 2010. However, it was mentioned that some aspects still need to be addressed. For example in Cocagne, NB, all participants mentioned the need to have an emergency shelter, which would also house a generator to ensure that people in need can go somewhere. In the winter of 2013, the fire station was able to take some 50 people, but this was the limit. As one

6.2 Results

person mentioned, his family ended up in another town as it was not possible to stay in their house after 48 h. They did not have a generator and when asked if he has one now he responded "No" and that they were still thinking about it. Two others of the same town had generators and felt they were less affected. Others who felt less affected were located on higher ground or at least a kilometer from the coast.

Preparedness has an important implication on how people will react to storms. The interviewees in general felt that they were still not really prepared for storms. As stated by one of the participants of Bonaventure: "*I don't know how to get ready for this.*" He continued by stating that when there is a forecast of a storm, he now brings everything from his backyard inside and places his car in a safe place far from trees (as he had damage from fallen trees in 2010). But at the same time, when asked whether he has tried to be more prepared, he replied: "*Not for now, I continue my daily routine.*" In Bonaventure, as in QC, where municipalities must have an emergency plan, it was felt that the community was prepared as much as it could knowing that they have problems regarding low floodplains. In NB, however, emergency plans were not available in all communities, as was the case in Cocagne. This explained why all participants expressed the need for an emergency shelter and generator.

Almost all of the interviewees mentioned that they were checking the weather forecast on a more regular basis since 2010. Most information was coming from the Internet, with some people referring also to television and on one occasion the radio. Environment Canada and Meteomedia were the most cited sources of information for weather forecasts. Communication was also mentioned as being more important and municipalities were doing better at this level. But there were many problems in some areas that needed to be addressed. In both NB and QC, half of the participants expressed the need to ensure that bylaws and regulations regarding the protection of the coastal zone be implemented. In Bonaventure, for example, it was mentioned that there should be a moratorium on building on the floodplain.

In addition to being more in-tune with the weather forecast, one person in NB said that he learned from the storm by buying large water jugs, as being on a well and needing electricity for this meant that during the power outage in 2013 there was no water for house operations (flushing toilets or drinking). By the end, all participants, except one, agreed that they had learned to be better prepared with small actions. One person from NB said: "*We learn with each storm.*"

The provincial governments in both NB and QC appeared to be ready in case of catastrophes. Public safety agencies were the most cited governmental organizations mentioned. All participants, to some degree, suggested that the agency was well organized and could respond relatively rapidly. One person in QC also stated that Hydro-Québec (the electricity company) was also fast to respond in case of a crisis. When asked about public health agencies, most did not understand their roles. Responses varied from "They are well-organized" to "Not hearing from them." One participant suggested that Public Health emits advisories when there are heat waves, while another person thought the agency was mainly for prevention.

Like in the first series of interviews in 2011–2012, 90% of participants stated that they did not ask for or receive help, except from a neighbor for a few of them. The provincial government helped in terms of roads blocked or flooded. However, one interviewee in QC was concerned about the lack of attention from the Ministry of Transportation regarding the blockage of the only road linking them to other cities. Instead of calling in, they advised Quebec Safety only. This left the town to try to find shelters for drivers in 23 cars that were stranded there. In NB, with the ice storm, three of the interviewees also mentioned the role of the Red Cross, which helped people who had no power and needed shelter.

When asked about the future and current challenges to be more adapted to climate change, the responses greatly varied. For the town of Cocagne, the main focus was on the development of an emergency plan and the purchase of a generator. This was mentioned by three participants. Three other participants from QC mentioned the need to protect beaches and shores, but this was still a challenge due to political pressure to develop. One person from NB also underlined the importance of having better coordination at the local scale, as it is not always possible to receive help from the province if roads are blocked or flooded.

At the personal level, each interviewee had a different answer. They varied from hoping to not live through another ice storm with a power outage again (interviewee in NB) to continuing to live as now. A person from NB explained: "*Our ancestors knew better; if you drive by all the old houses, [they] are all at about one kilometer from the coast or on high lands.*" He admitted that it is only in the past decades that people build at the coast. Finally, another person from QC declared that "*I will adapt to the temperature.*" He added, however, that those living at the coast will have to adapt to more flooding.

6.2.1 Knowledge of the CCC-CURA Project

Interestingly, when asked about their knowledge and level of participation in the project's activities, all except three persons mentioned not being involved. A few suggested that they had heard of the project and may have been in an information session, but did not remember if it was related directly to the project. One of the persons from NB said that he went to the winter institute that was organized in Moncton. He especially remembered the purple butterfly, the logo of the project, and a session on public engagement. The other two persons from QC vaguely remembered the activities and suggested that the project helped especially in terms of increasing awareness of people on the shore. As one participant mentioned, the main challenge is that if activities were not continuous and on a regular basis, people tended to forget.

6.2.2 Resilience

In the first set of interviews in 2011–2012, a question focused on the concept of resilience. Since the CCC-CURA project aimed to enhance the resilience of communities, this question aimed to understand whether people felt more resilient. Contrary to the 2011–2012 interviews, most participants declared they were resilient. However, resilience was defined differently by people. The words used included accepting variability, rebounding, and surviving. It was clear from the way that most interviewees explained resilience that they were talking about social resilience. One participant from NB suggested that fishers were proud and always exposed to unpredicted elements: *"There is nothing we can do against nature."* Another from QC said: *"I am not worried. We are capable to rebound."* Finally, another QC person stated: *"All the men here, almost, have a chainsaw and they like to go in woods, they are hunters. I'll say that if a catastrophe happens, they are less powerless than in the city. I think everybody would help, get their tools and rebuild."*

6.2.3 Natural Environment

Participants were asked whether they believed that the natural environment was prepared to deal with catastrophes. Responses were mixed, ranging from a complete affirmation to negation. A participant in QC suggested that the natural environment had no choice to adapt, while another one in NB stated that it was more prepared to deal with catastrophes than human communities. Half of the participants mentioned that, because of the consequences of human activities, humans may have no choice but to help the natural environment by either restoring or mitigating impacts. One of the examples given was the warming of the ocean and the introduction of exotic species, such as the green crab in the Northumberland Strait, leading to threats to the current commercial fisheries. One person underlined the importance of keeping the natural coastal environment: *"The marshes are there for reasons; it is to absorb large flux of water when the sea is high… Then when we block the marshes, this water has to go in residential development and this is in the houses."*

The aspect of restoring the coastal ecosystem was addressed by half of the participants. The most frequent suggestion was to restore the dunes and the shorelines. To do so, one participant stated that they used old Christmas trees and deposited them on the dunes so that sand can accumulate. However, he admitted that this was still only a short-term solution. Similarly, the same individual from NB mentioned that using rocks to protect the shore was not a good solution. It is a short-term "Band-Aid" solution and gradually would infiltrate in other ways. Another person from NB suggested that there was a need to diversify the species of trees that were planted. Most people planting trees tended to choose conifers, but with changing climate other species should also be promoted.

6.3 Discussion

The results of the second series of interviews bring a few key messages: (1) despite having experienced extreme events, people tend to revert to their old habits; (2) coastal communities in Atlantic Canada appear at least socially resilient; and (3) convincing people to adapt to climate change may be more complicated than expected. Changing attitudes of people and communities in regard to adaptation to climate change can be quite challenging. Burch et al. (2014) coined the term "path dependency" to describe the difficulty to change a system in order to enhance resilience or reduce greenhouse gas emissions. As stated, "path dependency is a key barrier, whereby alternatives become increasingly less likely over time as learning accumulates, irreversible choices are made, and inter-related elements of a system become deeply intertwined (Burch et al. 2014, p. 10). The main challenge with this barrier is that it reduces the capacity to innovate and find solutions that can be more sustainable. Baird et al. (2016) suggested that there is a pragmatist mindset when environmental issues, such as water governance, have to be addressed. In this study, we showed that pragmatism can occur at the individual or community level. This is reflected by how it has been difficult for most of these communities to not only develop an adaptation plan, but also implement it. Political will may not always be there. As mentioned by one of the interviewees, the short political cycle of 4 years does not help with implementing new solutions. A longer time horizon may be needed in politics in order to develop long-lasting strategies and have them implemented (Burch et al. 2014; Vasseur et al. 2017).

It was clear that most people in the communities that we studied relied mainly on their families or their neighbors when help was needed during an extreme event. These results were similar to those of the first interviews. Social capital represents the networks and resources present in a community and can be viewed through its level of social cohesion (Reininger et al. 2013). In recent years, there has been an increased interest to examine the importance of social capital in disaster preparedness. Reininger et al. (2013, p. 51) described that, "an individual's preparedness is reciprocally determined through such things as the amount of available material and intellectual resources (e.g., emergency funds and personal disaster kits; timely access to disaster alerts and knowledge of evacuation routes), their social support networks (e.g., families, churches, local response organizations), the community-level preparedness (e.g., relationships between emergency services, non-governmental organizations or NGOs, local businesses, community organizations) and the ability of the community to access and leverage resources from those in power (public officials, federal or international aid agencies)." Understanding the social capital present in a community can help to determine whether people feel capable of dealing with catastrophes. While this was not officially assessed in this study, we could link some factors of social capital to preparedness as discussed by Reininger et al. (2013). For example, those who felt more prepared to deal with storms were generally located further away from the coast and were older than those who mentioned being less prepared.

6.3 Discussion

Although this was not the focus of the analysis, it appeared that most people stating that they were prepared tended to also declare that they were resilient or at least more resilient than those who admitted to not being prepared. Individual resilience may be linked to the history of these families, as most of the participants have been living in these coastal communities for generations. One participant from NB, in fact, stated that: "*Resilience is like accepting. Best are fishermen. These people are proud. Here will continue to live like this.*" Rotarangi and Stephenson (2014) show that in a Maori community, cultural resilience can be strongly linked to social-ecological resilience and that it is not possible to omit culture when examining the capacity of a community like this one to adapt, transform, and improve its resilience capacity. We conceive that in coastal communities, having a specific culture that originates from the beginning of colonization in Atlantic Canada and their heavy reliance on fisheries, may also affect their conceptualization of resilience. However, the definition of resilience remains a challenge for most of the interviewed persons as it was at the beginning of the project (Vasseur et al. submitted).

At the community level, resilience can take another dimension, since it is related to its dynamics and socioeconomic capacity. The Community and Regional Resilience Institute defines community resilience as follows: "Community resilience is the capability to anticipate risk, limit impact, and bounce back rapidly through survival, adaptability, evolution, and growth in the face of turbulent change" (http://www.resilientus.org/about-us/what-is-community-resilience/). It was unclear in this study whether community resilience was seen as strong. Some elements, such as having an adaptation plan, acquiring a generator, and modifying the community center to become an emergency shelter when needed, could suggest that most of these communities were moving toward improving their resilience. However, this appears to be a process that may take time and can change, depending on socioeconomic factors and social configuration in the community. Eachus (2014) suggested that social cohesion is also an important ingredient in defining community resilience, as individuals interacting together through their local institutions and organizations bring the capacity of the community to mobilize.

It was interesting to see that participants had a wide diversity of responses regarding the resilience of ecosystems. The main message was that ecosystems and species within these ecosystems can adapt, but because of human activities, including climate change, there may be a need for support and help through restoration or management. In these communities that rely so much on natural resources, knowing that the ecosystem may be degraded can be a dangerous path for the future. In resource-based economies, the link between community and environmental resilience is very strong (Adger 2000). As being a part of this social-ecological system (SES), the communities did not necessarily realize their connections to the environment and it was taken for granted in their lives.

Community engagement and participatory processes in these study communities did not seem to be an issue, as people were connected through their own social networks and most stated that they were involved in the community. Within a community, most participants had similar views and knew the same challenges that they had to face (e.g. the need for a generator and an emergency shelter in one community of NB). The main challenge for these communities was to understand how to visualize the long-term consequences of these extreme events and how, in the future, they could be more proactive to better respond to changes. Lei et al. (2014) suggested that adaptation can be divided into two steps. (1) Short-term adjustment relates to responses to what had recently happened, while (2) long-term adaptation is based on lessons learned (considered in the next chapter) and innovations.

Increasing the awareness of people to climate change impacts and the need to adapt may help in the long term. However, the results showed that unless activities were conducted on a regular basis, people tended to forget relatively rapidly. There is a fine balance in this approach. One participant explained that he no longer watched some of the television shows related to climate change, as he felt that too much was pushed on people. Social acceptability of climate change adaptation and mitigation has been shown, therefore, presents another barrier to implement actions (Vasseur and Pickering 2015).

Few studies have taken a longitudinal approach to assess how communities perceive and work with the concepts of resilience, adaptation, and preparedness. Yabes and Goldstein (2015) reported that, in the northern Philippines, the capacity to adapt can increase over time when people are exposed to recurrent storms and other drastic environmental changes. In their case study, collaborative community governance has been a critical factor to enhance resilience, with learning networks allowing to gradually evolving as an SES. Considering that networks and social capital are important elements for communities to innovate and adapt, the learning process is, therefore, essential to open up new approaches. Long-term longitudinal studies can help to extract what types of ingredients can support the development of a resilience mindset in communities.

References

Adger WN (2000) Social and ecological resilience: are they related? Prog Hum Geog 24:347–364
Baird J, Plummer R, Bullock R, Dupont D, Heinmiller T, Jollineau M, Kubik W, Renzetti S, Vasseur L (2016) Contemporary water governance: navigating crisis response and institutional constraints through pragmatism. Water 8:224
Burch S, Shaw A, Dale A, Robinson J (2014) Triggering transformative change: a development path approach to climate change response in communities. Clim Policy 14(4):467–487
Burton I, May E (2004) The adaptation deficit in water resource management. IDS Bulleton 35:31–37
Eachus P (2014) Community resilience: is it greater than the sum of the parts of individual resilience? Proc Econ Financ 18:345–351
Lei Y, Wang J, Yue Y, Ahou H, Yin W (2014) Rethinking the relationships of vulnerability, resilience, and adaptation from a disaster risk perspective. Nat Hazards 70:609–627
Reininger BM, Rahbar MH, Lee M, Chen Z, Alam SR, Pope J, Adams B (2013) Social capital and disaster preparedness among low income Mexican Americans in a disaster prone area. Soc Sci Med 83:50–60

References

Rotarangi SJ, Stephenson J (2014) Resilience pivots: stability and identity in a social-ecological-cultural system. Ecol Soc 19(1):28

Vasseur L, Pickering G (2015) Feeding the social animal: how to engage Canadians in climate change mitigation. In: Potvin C (ed) Acting on climate change: extending the dialogue among Canadians, pp 163–170

Vasseur L, Horning D, Thornbush M, Cohen-Shacham E, Andrade A, Barrow E, Edwards S, Wit P, Jones M (2017) Complex problems and unchallenged solutions: bringing ecosystem governance to the forefront of the UN Sustainable Development Goals. Ambio. Accepted 4 Apr 2017

Vasseur L, Plante S, Znajda SK, Thornbush M (submitted) How coastal community members perceive resilience: a case from Canada's Atlantic coast. Ambio

Yabes R, Goldstein BE (2015) Collaborative resilience to episodic shocks and surprises: a very long-term case study of Zanjera irrigation in the Philippines 1979–2010. Soc Sci 4:469–498

Chapter 7
Implications and Lessons Learned

Abstract In order to improve the resilience of social-ecological systems of small rural coastal communities in Atlantic Canada, we used different methodologies to not only increase awareness and understanding of what is climate change, but also to enhance public engagement and find solutions that are more appropriate to these communities. This entailed, for example, integrating scientific and existing knowledge. Communities need to be a part of planning and governance in order to identify the most important impacts and response measures. Local governance, from local service districts to municipal and support from the provincial government, are also considered essential in order to improve resilience in the study region. It is necessary to have both short-term (emergency) plans in effect as well as longer term planning for improved adaptation to increasing major storms. Adaptation measures have to consider any physicosocial aspects of hazards, with social responses integrated into physical and environmental hazard-proofing initiatives. Other suggestions are included here based on lessons learned from the research.

Keywords Social-ecological resilience · Adaptation · Spatial scale · Temporal scale · Planning · Policy-making · Governance · Participative methods

7.1 Introduction

Historically, communities across the world have had to deal with environmental change and adjust to new conditions. Over centuries, people have learned to modify their infrastructure or habitat in order to adjust to the environment and better sustain their livelihoods. Today, some islands are disappearing as they are submerged by the sea. However, with population growth, migration in the coastal zones of the inland population, changes in socioeconomic activities, with most becoming more sedentary, and with greater complexity of the built infrastructure, communities have become less flexible in the way that they can respond and survive to changing conditions. This is especially true when conditions are changing rapidly.

© The Author(s) 2018
L. Vasseur et al., *Adaptation to Coastal Storms in Atlantic Canada*,
SpringerBriefs in Geography, DOI 10.1007/978-3-319-63492-0_7

Adapting to current anthropogenic climate change has been considered one of the most complex challenges for local communities, which are becoming more vulnerable to impacts over time if nothing is done. Since the last Intergovernmental Panel on Climate Change report (IPCC 2013), several new initiatives have examined the possible strategies for communities to adapt and many organizations across the world have been working to enhance the resilience of communities. However, outcomes remain limited and there are but few cases of long-term successful projects. One lesson learned from most of these case studies, shows the need to work at the local scale for more effective results to increase the resilience of communities (Vasseur et al. 2017). Newton and Weichselgartner (2014) have also underlined the importance to pay more attention to existing local knowledge (traditional and ecological) to better understand how people respond to changes.

This is especially true for coastal communities. Traditionally, coastal people have been accustomed to dealing with environmental and climatic elements. Their distance from urban centers, their relative isolation, and low density of their populations are other factors to consider for better understanding of territorial dynamics. In many cultures, mobility has been the norm for centuries, moving more inland during the harsh season and coming back to the coast during the summer or favorable times. For example, the Mi'kmaq communities of Atlantic Canada used to spend their summer along the coast for fishing and medicinal herb harvests and retreated into forested areas during the winter, when food could be more available through hunting (Vasseur and Tremblay 2014). Such communities were then considered resilient, although damage and loss of life were sometimes possible. However, with greater pressure on the coast due to urban development, mobility is more limited, affecting the capacity of people to respond to coastal changes (Bennett et al. 2015).

With a larger and less mobile population, more stable and permanent structures as well as less flexible socioeconomic systems, these communities face challenges as they can no longer afford to move or respond to changes rapidly. Decision making, planning, and management processes are complicated, more or less flexible, and often require a great deal of consultation before moving into action. In addition, the cost of most interventions is high and inaccessible for rural or coastal communities that are usually limited in their funding due to a low population size and limited revenues. This was very obvious in most of the communities that we studied. In many cases, through interviews and interactions, we learned that many could not afford to move their houses or to enhance protection using various technologies. Other participants did not feel the need to change, since extreme events are considered to be a normal way of life in these communities.

The main challenge for these communities is not only to make a decision, but how to make it in order to integrate the various aspects of adaptation and social-ecological systems (SESs) to improve resilience. Actors are faced with a problem related to climate or environmental change, but they do not know on what the decision should be based. While many continue to argue that decisions should be made solely on cost-benefit analysis, there needs to be a better understanding of other issues, such as social acceptability and capacity as well as governance. While there may be several guidelines and tools being developed in order to help communities adapt to climate change, most are complex and require expertise. Some of these tools may be available, but adaptation strategies remain vague and not necessarily adequate for a specific community. In this chapter, we discuss how the various components of the Coastal Community Challenges-Community-University Research Alliance (CCC-CURA) project attempted to integrate at the community level different tools and knowledge in order to co-produce adaptation plans and improve the resilience of coastal communities. The last section of this chapter extracts some of the lessons learned from this 6-year project.

7.2 Resilience to Climate Change

Resilience research has become popular since the late 2000s (Flood and Schechtman 2014) and is an ongoing area of research for fragile ecosystems, such as coastal systems. Approaches greatly vary among studies, but should arguably entail integrating social and ecological systems (Flood and Schechtman 2014; Holling 2001; Maldonado and Moreno-Sánchez 2014). Resilient communities in the context of climate change adaptation, for instance, are those capable of recovering or "bouncing back" following change, with the assumption that both human and ecological systems would remain functional. Alternatively, when we use a system approach, and if the system crosses its thresholds, a new state may be achieved. In such a case, the SES may be transformed and functional in what is often called a novel or "transformed" ecosystem.

Active adaptive management strategies are required when there is a regime shift that occurs reducing resilience (e.g., climate change) by affecting the magnitude, frequency, and/or duration of disturbances (Folke et al. 2004). Since humans are involved in an integrated systems approach, they need to be incorporated as part of a social dimension in resilience work and a vital part of SESs (cf. Folke 2006). "Social resilience" was addressed by Shaw et al. (2014) for the elderly experiencing coastal flooding in an assessment of their cognitive strategies and coping, which could be both individually and communally enhanced through preparation, or debilitated and result in "negative resilience" when there are misconceptions about the level of resilience.

7.3 Integrating Governance into Social-Ecological Resilience

According to Adger et al. (2005), in order to promote social resilience there must be "institutions for collective action, robust governance systems, and a diversity of livelihood choices are important assets for buffering the effects of extreme natural hazards and promoting social reorganization." These authors consider resilience before and after a disaster and advocated multilevel governance in order to enhance coping with uncertainty through mobilization. In particular, as demonstrated in their Table 1 (p. 1038), diverse ecological systems (also advocated by others, e.g. Folke et al. 1996, who relay the importance of biodiversity in order to promote resilience at the ecosystem scale; as part of "ecological resilience" in marine ecosystems, Hughes et al. 2005; Moore et al. 2009 conveying the significance of functional diversity; Cumming et al. 2013 considering ecosystem service provision) and economic livelihoods are needed. There is also a need for inclusive governance structures at the local scale in order to reduce vulnerability and boost adaptive capacity. In the midst of scarce financial provisions, countries are turning to adaptive governance involving community (local) participation (Schmidt et al. 2013). Celliers et al. (2013) referred to "cooperative governance" for integrated coastal management that adopts (national and international or global) policy by agencies at the level of the local government.

Multilevel governance is perhaps most instrumental to dealing with climate change adaptation because of the current hierarchical system operating as a baseline for new approaches (Cosens et al. 2014). Folke et al. (2010) recognized the possibility of smaller scale transformational change in order to promote resilience at larger scales. They also acknowledged the importance of adaptability (adaptive capacity) as part of resilience. It is possible to attain a greater adaptive capacity through interventions (e.g., policy, programs, actions, etc.) at various scales (Bennett et al. 2014). Sustainability itself has been defined (e.g., by Holling 2001, although not by everyone) in terms of creating, testing, and maintaining adaptive capacity. Whereas the transformability of a system denotes a new system, adaptability represents the ability of actors to influence (manage) resilience (Walker et al. 2004). Perceptions of transformation tend to focus on socioeconomic, political, and cultural aspects because they are vital to everyday life (Graybill 2013). This has been conveyed by Maldonado and Moreno-Sánchez (2014) as the three dimensions involving socioeconomic, social-ecological, and sociopolitical, with the social dimension as overarching and, hence, integrative (see their Fig. 2). These authors already identified indicators of these different dimensions, including (1) social-ecological: resource-use dependency, ecological awareness, and anticipation of disturbance; (2) socioeconomic: occupations, poverty, and infrastructure; and (3) sociopolitical and institutional: structural and cognitive social capital and perception of (marine) protected areas.

7.4 Coastal Community Resilience Planning

One of the first tools that was developed in this project was a toolkit to implement a dialogue on planning community resilience to environmental and climate changes (Vasseur 2012). The objective of this kit was to encourage and help communities implement a community dialogue to develop a consensus on which community

7.4 Coastal Community Resilience Planning

elements were vulnerable in their community to environmental and climate changes and from these vulnerabilities focused on strengthening the resilience of the community. It was based on the need for community dialogue to help first understand a specific issue and through it be able to look at potential solutions from different angles. So, actors were able to examine the ecological, economic, and social impacts of an issue (e.g., coastal erosion in town) to gradually discuss these potential strategies to reduce the impacts of this issue. Each solution was then analyzed in terms of feasibility, costs, technologies, social acceptability, etc. Once a consensus was reached, a solution or more could be implemented, monitored, and adjusted, if needed. Then, the next priority issue would be discussed. The kit was based on some basic principles including: inclusiveness, capacity-building, social acceptance, alternative and simple solutions, dialogue, openness and transparency, and complete understanding of the issues and solutions.

In the communities of Ste.-Flavie, Rivière-au-Tonnerre, Maria, and Bonaventure, this iterative kit was supported by various other tools that helped the dialogue among actors. These tools were integrated into the participatory action research (PAR) approach that was the basis of the project. These included, for example, the Method of Evaluation by Group Facilitation (MEGF) (Plante et al. 2016). To initiate the MEGF, "kitchen assemblies" (meetings held in neighborhood places) and focus groups were organized. These case studies have led to substantive results (e.g., participative mapping), such as a better understanding of the emerging issues (territorial planning, awareness/education, and ecological safeguarding), the development of community resilience plans, integration of hazards and vulnerabilities, identification of adaptation solutions, actions, and indicators to follow the progress made by the communities in terms of adaptation strategies and actions as well as resilience.

Assessing the work done in these communities, it was possible to extract some positive aspects that should be considered for future co-construction activities. It was felt that, in practice, the process was inclusive and participatory. People felt comfortable exchanging and discussing issues, as a common language was developed in the initial phase of the project in order to ensure that all people knew what the terms meant. Solutions were, therefore, reached through a consensus or voting process. It is expected that this type of approach would lend to greater appropriation and social acceptability. Indeed, Bell et al. (2013) have also reported that the approach is flexible enough that it is possible to review and reanalyze the data when conditions change. The other advantage of this method was its complementarity with other more traditional tools that were being used for municipal planning. However, the process was initially complicated to start at the time when the participants got full ownership of the various concepts used in the kits and understand how they could work and manipulate the tools to be more appropriate to their needs.

In other communities, other PAR approaches were used. For example, in Shippagan and LaMèque (NB), while similar group discussions occurred, visualization through maps was also used as a tool to increase understanding what may happen with storm surge and flooding. In Cocagne, Grand-Digue, and Dundas (NB), the PAR approach differed due to a main challenge. Indeed, some of these communities were not incorporated entities and, therefore, their preoccupations were somewhat different. Decision making being overseen at the provincial level, they wanted to mainly discuss how to implement at the local level some basic social movements that could help them to enhance resilience. A major step in Shippagan and Lamèque, for example, was to bring to the same table representatives from the various sectors of the communities and remove this idea of dealing with issues in isolation. With the communities being supported in the longer term, the risk of disengagement was significantly reduced. Having local data, because of another project being completed at the same time, helped to enhance understanding and maintain the engagement of people in the process. Transparency was not always easy to maintain, however, due to some political interferences. In most municipalities, where elected people are there on a 4-year electoral cycle, manipulations for future gains can be a challenge.

Other observations and lessons learned from the various activities in other communities included the importance to promote collective entrepreneurship and improve confidence and trust among actors. It is important to underline that coastal communities tend to be very proud of their lives and have a strong sense of ownership of their coastal areas. This was especially true for older people who had been there for generations. However, due to limited employment opportunities, younger people tend to migrate toward larger urban centers where jobs are more readily available. This out-migration posed a problem with most of the CCC-CURA study communities. This led municipalities to be limited in terms of finances and capacity to act on some of the issues. In the case of Ste.-Flavie, the trauma left behind from the December 2010 storms on individuals and the community was a challenge due to lack of trust. The need to plan differently in this community left people confused of how to redevelop the town and adapt to changes. For this reason, the work in this community took some time to bring results. One of the main challenges that was also pointed out in some communities was the degree of personal investment for some of the participants. This might not always be possible and, therefore, reduces the number of people involved. Where municipality support was available, however, it appeared that the process was moving somewhat faster. Nevertheless, financial support might have been circumscribed, thus, slowing the capacity to implement solutions.

7.5 Moving Further in Enhancing Resilience Through Ecosystem Governance

One of the main ambitions of the CCC-CURA project was to integrate the concept of governance into the resilience approach. Governance has taken many definitions in the past. This project initially defined governance according to Verbruggen (2007) and the Millennium Ecosystem Assessment (MEA 2005): the process of regulating behavior in accordance with shared objectives. It recognized that the

7.5 Moving Further in Enhancing Resilience Through Ecosystem Governance

contributions of various levels of government (global, national, regional, and local) and the roles of the private sector, nongovernmental actors, and civil society. During the course of the project, the concept evolved and developed into a more inclusive definition: interactions among structures, processes, and traditions that determine how responsibilities and power are shared in order to make decisions and how all actors have a say in ecosystem management. There were some common principles that communities held in common: (1) ecosystems are essential for sustainability; (2) the community must be consulted and involved in all steps; (3) it is necessary to integrate scientific and existing knowledge (e.g., cultural, ecological, and traditional); and (4) everyone must be open to everything as a way to adapt. The importance of natural ecosystems to help adapt to climate change hazards and maintain socioeconomic activities has been demonstrated by Spalding et al. (2014). The integration of scientific and existing knowledge has been underlined as positive in a number of studies (e.g., Aswani and Hamilton 2004; Fatoric and Morén-Alegret 2013).

Governance has been viewed from different angles in the project and in order to be considered "good governance" some ingredients are essential. The first one that was defined by the communities in their co-production of knowledge was the importance of dialogue. A second ingredient was how learning can be accessed and worked by the communities. There was a need to not only learn from experts, but also among themselves. The third ingredient targeted the process of decision making, which had to be open and transparent. This was probably one of the most fragile ingredients due to political conflicts in some of the communities. Having a long-term vision was felt as an ingredient necessary to ensure that everyone worked toward a common goal. Sustainable development was considered as part of the long-term vision that people should be keeping in mind during the process. Finally, most communities felt the need to integrate the notions of equality and equity to make sure that all cultures (Acadian, indigenous, and English), men and women as well as youth and the elderly were all included in the long-term vision.

In recent years, to acknowledge the need to encompass both social and ecological systems in making decisions regarding resilience and climate change adaptation, particularly in the view of sustainable development, the term "ecosystem governance" has been introduced (Vasseur 2016). Ecosystem governance requires respect for adaptive governance, ecosystem services, and biodiversity as a way to enhance social-ecological resilience. It needs to be bidirectional, connecting top-down national policies to bottom-up development strategies, and move from local to global through devolution (Vasseur et al. 2017). In order to accomplish this, the project would need to go further in defining ways to enhance the capacity of communities to act on their own turf regarding their view of sustainable development. This will also require research on new models that promote ecosystem services in decision making, policies to integrate SESs, capacity-building in all spheres and public engagement.

A similar process in the indigenous territory of Tanaca in Bolivia can serve as a good example of how communities can move toward more effective ecosystem governance (Painter et al. 2013). In this case, the territory faced issues of deforestation and loss of biodiversity as well as increased impacts due to climate change. The territory initiated a project using a participatory approach to first learn through climate change scenarios of what was to be expected in the future. Based on SES principles and dialogue in the community, they examined possible solutions in terms of afforestation of agricultural fields that were threatened by climate change and other ecosystem management options. Equity and inter-generational views were taken into consideration for a more sustainable future. In such a system, not only were mitigation and adaptation to climate change solutions defined and implemented, but also biodiversity conservation was enhanced through the maintenance of corridors for wildlife, sustainable subsistence hunting, and reduction of deforestation. They enacted policies to avoid the over-exploitation of resources, reduce deforestation and, therefore, landslides during heavy rainfall events, and define alternative (more sustainable) activities in the community. This type of ecosystem governance involving people and embracing integrated adaptive management has led to greater sustainability of the community. As stated by Painter et al. (2013): "I think that what the Tacana have done is more successful and sustainable than what is being done by their neighbors because their territorial management reduces the loss of forests and their biodiversity and ecosystem services, while also strengthening indigenous rights." Similar directions should be possible in coastal communities if they were to embrace a similar approach, which would go further than what was accomplished with the CCC-CURA project.

The lessons learned from this example and what we learned from the CCC-CURA project entail that it is essential to take into account all types of knowledge in considering SESs and the impacts of climate change. It is also important to understand the current type of governance and government existing in those communities and how to enhance it through more participatory approaches. Considering gender and inter-generational issues in decision making and the long-term vision of resilience of the community can lead to greater sustainability of the community. By the end, it was clear from this project that this is a long-term process of figuring out what will work best for each community, which necessitated reemphasizing the urgency to start talking about ecosystem governance for social-ecological governance. It also demonstrates that each community has its own approach to issues and the way that it engages in these processes. As stated by Bennett et al. (2015, p. 2): "Multiple socioeconomic and biophysical changes occurring simultaneously at different scales and speeds interact to produce drastically different outcomes for communities in different places."

7.6 Conclusion

Communities understand the need to adapt to climatic and environmental changes. The concept of resilience, however, may be more abstract and requires more understanding by communities in order to move in this direction. The present chapter briefly summarized how different tools could be used to help decision

7.6 Conclusion

makers and various local actors in the process of planning, decision, and management of their SESs in order to better adapt to climate change and improve resilience. With increased vulnerability levels in coastal communities, it becomes imperative that they understand their roles in these steps toward improving the resilience of their communities. Using a balanced approach to better visualize the conditions, communities can better appreciate how, over time and through careful and strategic planning, citizens can be able to gradually adjust and deal with changes. Through our longitudinal project, we predicted that communities would be able to help co-construct the tools, making it more accessible to others. While we acknowledged that there were differences between local communities, we also believed that common tools and processes could be useful for them to communicate and help each other. Coastal communities have a long tradition of helping each other. Under these conditions, and considering that many experience similar challenges, we believe that, over time, their capacity to co-construct will enhance the resilience of their communities as well as those of the communities in the region.

References

Adger WN, Hughes TP, Folke C, Carpenter SR, Rockström J (2005) Social-ecological resilience to coastal disasters. Science 309:1036–1039

Aswani S, Hamilton RJ (2004) Integrating indigenous ecological knowledge and customary sea tenure with marine and social science for conservation of bumphead parrotfish (*Bolbometopon muricatum*) in the Roviana Lagoon, Solomon Islands. Environ Conserv 31:69–83

Bell S, Correa Pana A, Prem M (2013) Imagine coastal sustainability. Ocean Coast Manage 83:39–51

Bennett NJ, Dearden P, Murray G, Kadfak A (2014) The capacity to adapt?: communities in a changing climate, environment, and economy on the northern Andaman coast of Thailand. Ecol Soc 19(2):5

Bennett NJ, Blythe J, Tyler S, Ban NC (2015) Communities and change in the anthropocene: understanding social-ecological vulnerability and planning adaptations to multiple interacting exposures. Reg Environ Change 16(4):907–926

Celliers L, Rosendo S, Coetzee I, Daniels G (2013) Pathways of integrated coastal management from national policy to local implementation: enabling climate change adaptation. Mar Policy 39:72–86

Cosens B, Gunderson L, Allen C, Benson MH (2014) Identifying legal, ecological and governance obstacles, and opportunities for adapting to climate change. Sustainability 6:2338–2356

Cumming GS, Olsson P, Chapin FS III, Holling CS (2013) Resilience, experimentation, and scale mismatchees in social-ecological landscapes. Landscape Ecol 28:1139–1150

Fatoric S, Morén-Alegret R (2013) Integrating local knowledge and perception for assessing vulnerability to climate change in economically dynamic coastal areas: the case of natural protected area Aiguamolls de l'Empordà, Spain. Ocean Coast Manage 85:90–102

Flood S, Schechtman J (2014) The rise of resilience: evolution of a new concept in coastal planning in Ireland and the US. Ocean Coast Manage 102:19–31

Folke C (2006) Resilience: the emergence of a perspective for social-ecological systems analyses. Global Environ Chang 16:253–267

Folke C, Holling CS, Perrings C (1996) Biological diversity, ecosystems, and the human scale. Ecol Appl 6(4):1018–1024

Folke C, Carpenter S, Walker B, Scheffer M, Elmqvist T, Gunderson L, Holling CS (2004) Regime shifts, resilience, and biodiversity in ecosystem management. Annu Rev Ecol Syst 35:557–581

Folke C, Carpenter SR, Walker B, Scheffer M, Chapin T, Rockström J (2010) Resilience thinking: integrating resilience, adaptability and transformability. Ecol Soc 15(4):20–28

Graybill JK (2013) Imagining resilience: situating perceptions and emotions about climate change on Kamchatka, Russia. GeoJournal 78:817–832

Holling CS (2001) Understanding the complexity of economic, social, ecological, and social systems. Ecosystems 4:390–405

Hughes TP, Bellwood DR, Folke C, Steneck RS, Wilson J (2005) New paradigms for supporting the resilience of marine ecosystems. Trends Ecol Evol 20(7):380–386

Intergovernmental Panel on Climate Change/IPCC (2013) Summary for policymakers. In: Stocker TF, Qin D, Plattner G-K, Tignor M, Allen SK, Boschung J, Nauels A, Xia Y, Bex V, Midgley PM (eds) Climate change 2013: the physical science basis. Contribution of working group i to the fifth assessment report of the intergovernmental panel on climate change. Cambridge University Press, Cambridge, United Kingdom and New York, NY, USA

Maldonado JH, Moreno-Sánchez RdP (2014) Estimating the adaptive capacity of local communities at marine protected areas in Latin America: a practical approach. Ecol Soc 19 (1):16–35

Millennium Ecosystem Assessment/MEA (2005) Ecosystems and Human Well-Being: Volume 2. Scenarios. Island Press, Washington, DC

Moore SA, Wallington TJ, Hobbs RJ, Ehrlich PR, Holling CS, Levin S, Lindenmayer D, Pahl-Wostl C, Possingham H, Turner MG, Westoby M (2009) Diversity in current ecological thinking: implications for environment management. Environ Manage 43:17–27

Newton A, Weichselgartner J (2014) Hotspots of coastal vulnerability: a DPSIR analysis to find societal pathways and responses. Estuar Coast Shelf S 140:123–133

Painter L, Siles TM, Reinaga A, Wallace R (2013) Escenarios de deforestacion en el gran paisaje Madidi-Tambopata. Consejo Indígena del Pueblo Tacana (CIPTA) and Wildlife Conservation Society (WCS), La Paz, Bolivia, 40 pp

Plante S, Vasseur L, da Cunha C (2016) Chapter 4. Adaptation to climate change and Participatory Action Research (PAR): lessons from municipalities in Quebec, Canada. In: Knieling J (ed) Climate Adaptation Governance. Theory, Concepts and Praxis in Cities and Regions. Wiley-VCH Verlag GmbH & Co. KGaA, Weinheim, pp 69–88

Schmidt L, Prista P, Saraiva T, O'Riordan T, Gomes C (2013) Adapting governance for coastal change in Portugal. Land Use Policy 31:314–325

Shaw D, Scully J, Hart T (2014) The paradox of social resilience: how cognitive strategies and coping mechanisms attenuate and accentuate resilience. Global Environ Chang 25:194–203

Spalding MD, Ruffo S, Lacambra C, Meliane I, Hale LZ, Sheppard CC, Beck MW (2014) The role of ecosystems in coastal protection: adapting to climate change and coastal hazards. Ocean Coast Manage 90:50–57

Vasseur L (2012) Getting Started with community resilience planning. A kit to implement dialogue on planning community resilience to environmental and climate changes. Training manual. 21 pages (23 pages in French). Produced for coastal communities challenges (SSHRC-CURA), Southern Gulf of St. Lawrence Coalition for Sustainability and the Regional Adaptation Collaboratives (RAC) climate change program

Vasseur L (2016) Looking at ecosystem governance as a way to achieve environmental sustainability and tackling "wicked" problems. In: Kurissery S, Turvey R, Pendea IF (eds) Environmental sustainability. Wiley Publisher, pp 114–125

Vasseur L, Tremblay E (2014) Coastal ecosystem in Kouchibouguac National Park of Canada: adaptation possibilities for protecting traditional knowledge of local a community. In: Buyck C (ed) Safe havens: protected areas for disaster risk reduction and climate change adaptation. IUCN, Gland, Switzerland, pp 33–40

References

Vasseur L, Horning D, Thornbush M, Cohen-Shacham E, Andrade A, Barrow E, Edwards S, Wit P, Jones M (2017) Complex problems and unchallenged solutions: bringing ecosystem governance to the forefront of the UN Sustainable Development Goals. Ambio (in press)

Verbruggen A (2007) Glossary. In: Metz B, Davidson OR, Bosch PR, Dave R, Meyer LA (eds) Mitigation of climate change. Contribution of working group III to the fourth assessment report of the intergovernmental panel on climate change. Cambridge University Press, Cambridge, UK and New York, NY, USA

Walker B, Holling CS, Carpenter SR, Kinzig A (2004) Resilience, adaptability and transformability in social-ecological systems. Ecol Soc 9(2):5–13

Chapter 8
Conclusions

Abstract In this final chapter, a summary of the overall findings across the years of this study is presented along with a discussion of its main conclusions and contributions. Among these, has been a general sentiment among participants of enhanced (social) resilience and improved emergency action. Although the short-term response is somewhat better, work is still needed to grasp any opportunities to benefit the longer term response and results when working with small rural coastal communities, such as those presented for Atlantic Canada.

Keywords Response · Short-term (emergency) action · Longer term action · Adaptation · Social resilience · Sustainability

This brief has delineated research following the 2010 winter storms in Atlantic Canada based on a participatory approach and interviews of individuals and couples in order to gauge individual-to-community level impacts and responses. Ten rural coastal communities located in the provinces of Québec (QC), New Brunswick (NB), and Prince Edward Island (PEI) were investigated, with a revisiting of some of these communities (located in QC and NB) subsequently in 2014. These communities were outlined in Chap. 4 and the results of the initial interviews held in 2011–2012 conveyed in Chap. 5, with the final interviews from 2014 outlined in Chap. 6 of this brief. Much has been learned from this research in working with the communities and following up on their experiences and this information, along with the implications of the findings, were expressed in the previous chapter. In the current final chapter, the authors will briefly summarize the overall findings and main conclusions derived from this work; they impart the contributions that it makes and suggest areas where additional work may be needed.

8.1 Summary of the Overall Findings

Even though much of the research is not yet published and this process is currently ongoing, some aspects of its findings can already be gleaned from some published articles as well as results from the analysis. The overall findings that can be derived from this study, including interviews held in 2011–2012 and 2014, are as follows:

- Impacts to personal property were identified in addition to more social-level effects to businesses and infrastructure that affected evacuation and emergency relief.
- Emotional reactions of fear, stress or worry, panic, and powerlessness were identified associated with stressors and damage, especially from those who had been cut off from the mainland and the elderly as well as those who could not afford to recover at the household level without any government support.
- There is an indication of the impact of demographics on community experiences and responses, and the variables, such as gender, age, occupation, and level of education, need consideration.
- Men and women experienced similar storm impacts, particularly flooding; however, women referred to domestic incidences, whereas men mentioned the flooding of roads and public infrastructure. This gendered experience (that was contingent on exposure) was affected by sex-typical occupations.
- The tendency to forget past storms (event memory) came across clearly in this research, as with respondents in 2014 tending to forget about organized meetings. It is suggested that because people tend to forget rapidly, activities need to be conducted on a regular basis.
- Local people were able to identify changes in weather patterns, such as higher summer temperatures and milder winters with less snow and more rain, an earlier spring, and more high tides.
- People have learned some lessons, including the need to keep a generator; not to leave important belongings in the basement; modify the height (location) of electric outlets; etc.; and being more prepared for storms.
- There has been some behavioral changes, as for example checking the weather forecast for information.
- There have been some improvements, including that of communication after the 2010 winter storms, emergency plans, faster emergency services (particularly with the recent ice storm in the winter of 2013), and a greater presence by the provincial government (in NB).
- The clientele's approach of the QC government during the December storms has created some frustrations between members of the communities.
- Often government measures (like financial help decree, fuzzy reglementation, and normative approach) can have side effects in terms of devitalization factors of the community.
- The importance of the collaboration among all actors involved in this topic and the fragility of processes caused by non-continuity action linked to political changes at the local level can be a barrier to advance the adaptation agenda.

8.1 Summary of the Overall Findings

- There are still some issues, however, that need address, such as the request for an emergency shelter in Cocagne (NB) and a generator.
- Restrictions are needed to prevent building at the coast, which has been instated only in the last few decades and become a problem.
- An individual reflex in front of the effect of storms is to search the symbolic strength with solid infrastructure, but the desire to share other examples of soft adaptation is mentioned.
- People recognized the need for adaptation, as to increased temperatures as well as flooding.
- In 2014, there was a general feeling of social resilience, in particular, among community members.
- Participants suggested restoring dunes and the shoreline as part of remediation, as through the use of conifers for stabilization and buildup.
- Those who felt unaffected were located on higher ground or at least a kilometer from the coast.

8.2 Main Take Home Messages

Even though there may not be an accurate understanding of terminology, such as "coastal zone" and "resilience," community members are able to comprehend that measures are needed in order to improve local conditions and improve their adaptive capacity. Suggestions, such as the need for flood protection (hazard-proofing), were presented during the interviews and this implied that participants were aware of what was required to reduce their vulnerability as small rural coastal communities. Using the participatory action research (PAR) approach, it was possible to discern their understanding based on exposure and past experiences. In the interviews, for instance, they were able to convey their suggestions for action and response to climate change and this indicated some level of locally informed contribution on the part of study participants. Importantly, this approach also enabled for an understanding of individual-couple experiences and impacts and responses to various climate change impacts, including storm surge, flooding, high waves, and coastal erosion. Nevertheless, there is a pressing challenge that exists for climate change adaptation in Atlantic Canada. Specifically, because communities are perceived to be socially resilient (in 2014 compared to 2011–2012), convincing people to adapt to climate change will be difficult also because of their tendency to revert back to old habits and perhaps forget experiences and thereby the lessons that can be gleaned from them.

"Good governance" in this region requires the co-production of knowledge through dialogue. Social learning through the exchange of information between communities has also been implicated to be vital, although it can be limited where human error is concerned (Wu et al. 2014). Decision making needs to be transparent, so that everyone feels that they are working toward a common goal and part of the vision of sustainable development. It is important to work with people of different sectors, who have an interest to engage themselves and who are ready to do something now. Local organizations need to be more integrated as well as well-run and accountable (Sultana and Thompson 2010), like certain environmental nongovernmental organizations have done in some places. Inclusion is essential and the integration of multiple actors is necessary when achieving a common voice and carrying out community action; this includes, for example, equitable access to local decision making by all social groups (Ahammad et al. 2014). Furthermore, it is important that an integrated approach, as provided by social-ecological systems (SESs), is encouraged in order to reach resilience and promote climate change adaptation. Researchers need to understand the actor network and existing governance systems in communities from the outset that may affect decision making and the long-term vision of these communities. The notion of ecosystem governance, for instance, will need to be integrated into already existing frameworks of landscape management and conservation.

8.3 Contributions

Although communities had not changed much in response to research initiatives in the span of this research, there have been some positive outcomes and contributions of the work. In itself, the organized meetings informed the public about some terminology and anticipated climatic and environmental changes. This promoted learning among individuals attending the meetings and presented a new framework for approaching the problems. In aged communities, where change is often difficult, it is not surprising that some people forget about interventions performed by the project to benefit these communities at risk. People quickly forgot past experiences and could not derive any great amount of lessons learned, making them less tenable to change. Nevertheless, there was a sense among interviewees between 2011–2012 and 2014 that some improvements were evident. For instance, respondents thought that communications were better and that emergency planning had also been ameliorated in some places. So, there were some positive outlooks in regard to municipal management activities and infrastructure, internalized lessons in strategic planning, preparation, and emergency response in particular.

Importantly, the role of social learning and behavioral change was confirmed. People are looking up the weather on a regular basis and so are more attuned to weather-related changes and possible storm activity. This could lubricate the emergency response (short-term adaptation) and lead to improved evacuation and reduced short-term impacts, including casualties and loss of life. People are also acting in some capacity to alleviate the effects of storms by contributing to beach stabilization efforts through conifer armoring of beaches and dunes. This suggests some levels of social resilience that is also communicated through the relocation of residents onto higher ground or further from the coast has been noted in this brief. Community members are also being more proactive in terms of demanding that the government restrict building at the coast (as through the call for reduced allocation of building permits and the acknowledgment that developing on wetlands induces reduced wetland storage of floodwater); and they want to be involved in decisions affecting the relocation and the attribution of vacant lots caused by the destruction of houses. An important word often mentioned during the PAR process is how to develop a sense of "social responsibility" or responsibility of the citizen.

8.4 Further Work

More work is needed to improve the long-term response to winter storms and their impacts, as for instance as regards the emergency response (shelters and generators). The government needs to ensure that residents have necessary structures and equipment in place as well as emergency action plans and evacuation procedures that are recognized by residents and with which whose execution residents are familiar. Perhaps yearly drills or training sessions would benefit new and older community members alike. In particular, new residents need to be made aware of the risks associated with coastal living and be familiar with the impacts and protocols, especially if they are not accustomed to being at the coast. More sensible low-cost approaches are necessary to accommodate these communities. For instance, perhaps there needs to be a restriction on building at the coast plus a regulation of mobile home use within a kilometer of the shoreline. There is already an existing mobility of residents, as some travel between their summer and winter properties in order to control exposure to storms and other seasonal hazards associated with extreme weather and climate change. This notion of increased mobility should perhaps be strengthened and further encouraged.

Although weather-monitoring systems are in place and are conveyed by Environment Canada and meteorological networks, the communication of warnings could be improved with better communication links (telephone lines and Internet) in all communities. In particular, those communities that are isolated from the coast and dependent on single throughways for navigation, need to be especially alerted to forecasts and warnings on a regular basis. They need to ensure that an emergency plan is in place and an emergency protocol, which includes roadway and infrastructural clearance, is effective so that their escape routes are not blocked during an extreme event.

When working with any community, it is necessary that case-specific knowledge is gained regarding decision making and governance as well as approaches. In addition to this, workers should be cognizant of community demographics as an indicator of composition and response. It has been relevant among the small rural coastal communities in this study that these are aged communities comprising mostly men with traditional occupations, level of literacy, and gender-specific roles that affect gendered experiences and reactions. This situation is particularly true in NB, but needs to be nuanced in QC, where a larger diversity of occupational activity was denoted. Knowing this may suggest that change is likely to be slow and dependent on authoritative routes, as by governments, and that change involves long-term engagement with these communities as well as ongoing or continuous drilling. This acts to alert new and aged residents to the change that is affecting their environments at the coast and that will, in turn, affect their communities and households down to the individual level. It is crucial that this action be sustained so that those with a tendency to forget or not welcome the novel or innovative have a chance to at least understand the need for intervention so that vulnerable communities at risk can at least have the opportunity to adjust to climate change and adapt.

8.5 Discussion

There are still constraints operating along shorelines that could circumscribe the adaptation process. For the small rural coastal communities in this study, among these are problems of financing major efforts to hold the coastline and the lack of clear objectives from state agencies. Individuals alone cannot fund major hard defenses, although some do manage to put up retaining walls to protect their own properties. In the early set of interviews, in the absence of private insurance, there was a call for help from the government to finance coastal defense (hard engineering) and spur the insurance companies to reimburse more than half of damage payments to households. In the more recent interviews, however, this approach seems to have been muted and people were more aware of soft defenses that would be cheaper to emplace, such as the example of the Christmas trees to armor the coast and encourage sand accumulation on beaches and to stabilize dunes, or experiences sharing in Ste.-Flavie with the re-vegetation activities documented (e.g., rules, principle, species). This is an instance where people are working with the shoreline to mitigate and remediate the effects of storms, storm surge, high waves, flooding, and coastal erosion.

Other authors (e.g., McNamara et al. 2011) similarly noted situations where human actions, as through shoreline nourishment, are being executed in order to combat erosion. They have stipulated that this interaction of people at the coast that is evident at the regional level depends on property value and the economic viability (so, capacity) of programs, such as nourishment. Large countries with long shorelines to protect (as with nourishment) will have to endure the most costs and small island states will be most affected by erosion (Hinkel et al. 2013). In this case study, particularly in 2014, support for cost-benefit analysis was still evident; however, a better understanding of issues (social acceptability and capacity as well as governance and resilience) needs consideration so that cost-benefit analysis is not the sole approach taken in the decision-making process. There needs to be a unified and well-orchestrated effort by people and government regarding concerted action at the coast. However, economics and social cohesion may be limiting factors or success factors. In other words, an integrated approach (social, political, economic

8.5 Discussion

plus environmental) is required for sustainability at the coast and the future of these communities.

In addition, it has been argued that initiatives must be place-based, innovative, and flexible (Craig and Ruhl 2010). This is a challenge for social acceptability perhaps, given aged communities that are set in their ways and unaccommodating of change and innovation and thereby affecting "social innovation" (Romida-Taylor 2012) and ability to learn to deal with uncertainties. Migration, for instance, can act as a tool to spread innovation across regions through knowledge transfer, technology, etc. as mechanisms (cf. Scheffran et al. 2012). However, it is encouraged that the response be from a place-based perspective, so as not to omit any local features, such as that of local decision-making strategies, culture, etc. It is also noteworthy that technical adaptation measures, such as for flood protection, are inflexible in comparison to collaborative planning at the regional scale that involved scientists working with communities in a participatory approach (Bormann et al. 2012).

Co-management strategies have been shown to be effective, as they also adopt a participatory approach to adaptation and operate using a site-specific approach that is integrated and involves local communities (Schmitt et al. 2013). Using this combination, it was possible to establish bamboo breakwaters in the mangroves of Vietnam, for instance, to counteract erosion and at the same time stimulate sedimentation. This is similar to the Christmas trees example acquired from the second set of interviews in 2014, when locals promoted such soft strategies, even though they were recognized as temporary or short-term measures, also including other trees beside conifers (diverse tree types) or sand rye grass (sea lyme grass) to control erosion and promote sedimentation along the coast.

There was some opposition to hard defenses in this study, and they were seen by some as ineffective temporary measures that are not (economically) sustainable. This finding has been communicated also by van Slobbe et al. (2013), who find traditional engineering, that typically ensures safety, to be suboptimal in other ways and ascertain to promote neither resilience nor sustainability. Evidently, when faced with limited funding, communities will prefer more economic approaches to coastal protection. Similar results are reported by Fatorić et al. (2014), who compare responses in Greece versus Spain to discover that half of actors prefer natural adaptation measures, as for instance sand dune and beach barriers set up to protect coastal wetlands. It should also be noted that these authors (Fatorić et al. 2014) also find a greater openness for emigration by Spanish actors involved in comparison to those in Greece. This indicates that relocation strategies are not well-suited to everyone and a site-specific approach be adopted, particularly where some community members are similarly less tenable to relocation as an adaptation measure. Temporary relocation, for instance, is already practiced in the study area and has also been discovered in other research (e.g., Motsholapheko et al. 2011) as an existing coping strategy.

In terms of the cost associated with flood protection, economic loss assuming no defenses should be considered alongside costs associated with flood impacts. Authors have cautioned about adaptation efforts necessary to overcome flood impacts, where the economic damage is reduced by 67% in cases where fewer people (down by 37%) are affected by the 100-year flood (Mokrech et al. 2015). So, doing nothing is only an option if relocation is employed and managed retreat (or "managed withdrawal" cf. Parkinson 2009) deployed to ensure that fewer people are actually exposed to coastal hazards and impacts. Where tourism has encouraged coastal development, coastal setbacks have been suggested in order to deal with coastal erosion, as in Costa Brava Bays, Spain and the Danube Delta, Romania (Sanò et al. 2011). Other studies of risk mitigation for coastal erosion and flooding have advocated the relocation of communities (e.g., Milligan et al. 2009; Ruckelshaus et al. 2013). This is especially pertinent if we are to consider that the situation at the coast could actually get worse and that we should prepare for larger future disasters, particularly those affecting coastal cities (cf. Hallegatte et al. 2013).

Since flooding worsens with urbanization and the paving of surfaces, it is advisable that restoration of the natural environment might work to counteract these effects. Vulnerable communities are further at risk if unprepared, especially when dealing with greater magnitude major events in the future. For this reason, it has been suggested that coastal salt marsh restoration, for example, be instigated in order to adapt to sea-level rise and mitigate climate change in the Bay of Fundy, Canada (Byers and Chmura 2007; Singh et al. 2007), including the potential for sea-level rise of ± 50 m (evident in Pleistocene glacial cycles, Woodruff 2010). Sea-level rise causing a storm surge of 0.5 m by 2080 could affect 47% more people and lead to 73% more property loss, as for instance in the southern shores of Long Island, NY (Shepard et al. 2012). Research has also shown that marshes protect shorelines from erosion, as during the Category 1 storm Irene that impounded the central Outer Bank, NC (Gittman et al. 2014). The protective element of marshes is also conveyed in the current study, particularly in 2014, when a participant observed that covering marshes was detrimental because they took up floodwater that would go into (and flood) buildings with development in these natural areas. This means that there is potential for "retreat strategies" (Nordstrom and Jackson 2013), so that developed sites revert back to natural processes. Ultimately, however, it may be necessary to retreat from coastal settlements (Craig 2014).

There is now a better understanding of the choices involving the response to sea-level rise (cf. Nicholls and Tol 2006). However, we are also coming to grip with the restrictions surrounding each option available. There are tradeoffs to consider, for instance, where shoreline protection that adopts large-scale structures, such as dikes, is limited by funds, whereas ecosystem migrations associated with retreat options may be constrained by land use (Hecht 2009) as well as political boundaries. For that, mapping risks and participative mapping about effective risks zones or perceived risks zones by the actors is necessary for the decision maker and citizens. However, follow-up with them is a prerequisite in order to be able to understand these results (some research groups are working on it like Bernatchez's team at UQAR University). Cooper and Pile (2014) have presented a dichotomy associated with adaptation to climate change, where, on the one hand, (1) human activities can be altered to accommodate environmental change (e.g., hazard-proofing, relocation, land-use change, etc.) and, on the other hand, (2) there is the possibility of resisting environmental change in order to accommodate current

8.5 Discussion

activities and infrastructure (e.g., flood defenses, including seawalls, beach nourishment, etc.). The authors report that the former (preserving human activities and infrastructure, which they termed "resistance" or "daunting") option is most prevalent over the "adaptation" option (involving change). They stressed that, whereas resistance is damaging to the environment, costly, and risky, adaptation offers the potential to reach sustainable outcomes in the longer term, although they may be politically challenging to implement.

It is known that social change takes time. This is, therefore, also true for the social dimension of adaptation (Hurlimann et al. 2014). In the current study, it became clear that change would take time; as for instance people's recognition of the information available to them via organized meetings and their perception of change that affected social learning. After all, it is necessary to establish priorities before the transfer of lessons (learning from mistakes) may occur (Tompkins 2005). As recently recognized by Serrao-Neumann et al. (2015), a transdisciplinary and inter-sectoral approach may be most appropriate to address such complex problems, as those emerging from climate change adaptation, where a dynamic learning approach, such as learning-by-doing or doing-by-learning, is necessary because of its flexibility. Coastal managers require learning opportunities as well in order to be able to effectively deal with climate change risks (Tribbia and Moser 2008). Others (e.g., Khan et al. 2012) have stressed the importance of communication and education in order to implement understanding informally through a community-based participatory approach that creates awareness, particularly among resource-dependent communities as part of a pivotal social perspective. Although this approach may be challenging, it represents a soft adaptation strategy that is based on enhancing human and social capital that is used to increase adaptive capacity and develop resilience (cf. Uy et al. 2011). For example, raising public awareness of oyster decline could be beneficial to stewardship (Scyphers et al. 2014). Such an approach to resilience is flexible, proactive, and more accommodating of local situations than traditional (top-down) approaches (Wardekker et al. 2010). Social factors have been considered to be more important than technical ones in adaptation and sustainability research (e.g., Marshall et al. 2011; Thornbush et al. 2013). Moreover, as noted by Rawlani and Sovacool (2011), technology (or a "technical or technological fix") is only one component of successful adaptation.

Demographic change that involves aging populations, as evident in the study area, has potential to augment the vulnerability of communities to the risks associated with climate change by affecting socioeconomic trends (Roiko et al. 2012). These authors have noted that in addition to these (aged) communities, indigenous people, lone-person households, and single-parent families are similarly affected as the aged; and their challenges are expected to worsen as populations grow and there is greater resource scarcity and competition. Poverty is another factor that influences small rural communities and its reduction requires a local and integrated framework (cf. Sales Jr. 2009).

More complex systems are also affecting governance to a point that multilevel adaptation has been conceived as insufficient on its own to deal with multiscale and multisector issues (Fidelman et al. 2013). In particular, these authors have noted the threat posed by short-sighted (short-term) adaptation measures that include incremental, sectoral, top-down approaches as well as interactions emerging from complex systems. As espoused previously in this brief (see Chap. 7), environmental sustainability requires long-term planning and policies that extend beyond electoral cycles. So, there is a conflict between short-sighted political practice and the temporal scope that is required for sustained environmental management.

Overcoming the complexity of systems and interactions that may pose problems with coping, planning, and policy-making is the challenge of interdisciplinary collaboration that can work to benefit sustainability (McMichael et al. 2003), including for instance linked geomorphological and ecological dynamics (Rhoads et al. 1999), but may be circumscribed by economic, political, and governance barriers. For example, the struggle to control development at the coast, which can be achieved by restricting permits, and zoning, but that remains elusive due to income generated from permit applications noted by participants in 2011–2012. There are also problems associated with increasing economic disparities emerging from capitalist systems that foster exploitation (as by local elites) and poor local leadership in some countries, as is evident in Kenya (Okello et al. 2009). This complicates the ability of equity and social inclusion, which may not be as great a problem affecting small rural coastal communities, and may be more relevant a consideration for expanding cities along the coastal zone around the world.

References

Ahammad R, Hossain MK, Husnain P (2014) Governance of forest conservation and co-benefits for Bangladesh under changing climate. J Forest Res 25(1):29–36

Bormann H, Ahlhorn F, Klenke T (2012) Adaptation of water management to regional climate change in a coastal region—hydrological change vs. community perception and strategies. J Hydrol 454–455:64–75

Byers SE, Chmura GL (2007) Salt marsh vegetation recovery on the Bay of Fundy. Estuar Coasts 30(5):869–877

Cooper JAG, Pile J (2014) The adaptation-resistance spectrum: a classification of contemporary adaptation approaches to climate-related coastal change. Ocean Coast Manage 94:90–98

Craig RK (2014) Using a public health perspective to insulate land use-related coastal climate change adaptation measures from constitutional takings challenges. Plan Environ Law 66(5):4–7

Craig RK, Ruhl JB (2010) Governing for sustainable coasts: complexity, climate change, and coastal ecosystem protection. Sustainability 2:1361–1388

Fatorić S, Morén-Alegret R, Kasimis C (2014) Exploring climate change effects in Euro-Mediterranean protected coastal wetlands: the cases of Aiguamolls de l'Empordà, Spain and Kotychi-Strofylia. Greece. Int J Sust Dev World 21(4):346–360

Fidelman PIJ, Leitch AM, Nelson DR (2013) Unpacking multilevel adaptation to climate change in the Great Barrier Reef, Australia. Global Environ Chang 23:800–812

Gittman RK, Popowich AM, Bruno JF, Peterson CH (2014) Marshes with and without sills protect estuarine shorelines from erosion better than bulkheads during a Category 1 hurricane. Ocean Coast Manage 102:94–102

Hallegatte S, Green C, Nicholls RJ, Corfee-Morlot J (2013) Future flood losses in major coastal cities. Nat Clim Change 3:802

References

Hecht AD (2009) The tipping points of sea level rise. Environ Res Lett 4:1–2

Hinkel J, Nicholls RJ, Tol RSJ, Wang ZB, Hamilton JM, Boot G, Vafeidis AT, McFadden L, Ganopolski A, Klein RJT (2013) A global analysis of erosion of sandy beaches and sea-level rise: an application of DIVA. Glob Planet Change 111:150–158

Hurlimann A, Barnett J, Fincher R, Osbaldiston N, Mortreux C, Graham S (2014) Urban planning and sustainable adaptation to sea-level rise. Landsc Urban Plan 126:84–93

Khan AS, Ramachandran A, Usha N, Aram IA, Selvam V (2012) Rising sea and threatened mangroves: a case study on stakeholders, engagement in climate change communication and non-formal education. Int J Sust Dev World 19(4):330–338

Marshall NA, Gordon IJ, Ash AJ (2011) The reluctance of resource-users to adopt seasonal climate forecasts to enhance resilience to climate variability on the rangelands. Clim Change 107:529–551

McMichael AJ, Butler CD, Folke C (2003) New visions for addressing sustainability. Science 302:1919

McNamara DE, Murray B, Smith MD (2011) Coastal sustainability depends on how economic and coastline responses to climate change affect each other. Geophys Res Lett 38:L07401

Milligan J, O'Riordan T, Nicholson-Cole SA, Watkinson AR (2009) Nature conservation for future sustainable shorelines: lessons from seeking to involve the public. Land Use Policy 26:203–213

Mokrech M, Kebede AS, Nicholls RJ, Wimmer F, Feyen L (2015) An integrated approach for assessing flood impacts due to future climate and socio-economic conditions and the scope of adaptation in Europe. Clim Change 128:245–260

Motsholapheko MR, Kgathi DL, Vanderpost C (2011) Rural livelihoods and household adaptation to extreme flooding in the Okavango Delta, Botswana. Phys Chem Earth 36:984–995

Nicholls RJ, Tol RSJ (2006) Impacts and responses to sea-level rise: a global analysis of the SRES scenarios over the twenty-first century. Phil Trans R Soc A 364:1073–1095

Nordstrom KF, Jackson NL (2013) Removing shore protection structures to facilitate migration of landforms and habitats on the bayside of a barrier spit. Geomorphology 199:179–191

Okello MM, Seno SKO, Nthiga RW (2009) Reconciling people's livelihoods and environmental conservation in the rural landscapes in Kenya: opportunities and challenge in the Amboseli landscapes. Nat Resour Forum 33:123–133

Parkinson RW (2009) Adapting to rising sea level: a Florida perspective. In: Nelson GL, Hronszky I (eds), Sustainability 2009: The Next Horizon. American Institute of Physics, pp 19–25

Rawlani AK, Sovacool BK (2011) Building responsiveness to climate change through community based adaptation in Bangladesh. Mitig Adapt Strateg Glob Change 16:845–863

Rhoads BL, Wilson D, Urban M, Herricks EE (1999) Interaction between scientists and nonscientists in community-based watershed management: emergence of the concept of stream naturalization. Environ Manage 24(3):297–308

Roiko A, Mangoyana RB, McFallan S, Carter RW, Oliver J, Smith TF (2012) Socio-economic trends and climate change adaptation: the case of South East Queensland. Australasian J Environ Manage 19(1):35–50

Romida-Taylor D (2012) Social innovation and climate adaptation: local collective action in diversifying Tanzania. Appl Geogr 33:128–134

Ruckelshaus M, Doney SC, Galindo HM, Barry JP, Chan F, Duffy JE, English CA, Gaines SD, Grebmeier JM, Hollowed AB, Knowlton N, Polovina J, Rabalais NN, Sydeman WJ, Talley LD (2013) Securing ocean benefits for society in the face of climate change. Mar Policy 40:154–159

Sales RFM Jr (2009) Vulnerability and adaptation of coastal communities to climate variability and sea-level rise: their implications for integrated coastal management in Cavite City, Philippines. Ocean Coast Manage 52:395–404

Sanò M, Jiménez JA, Medina R, Stanica A, Sanchez-Arcilla A, Trumbic I (2011) The role of coastal setbacks in the context of coastal erosion and climate change. Ocean Coast Manage 54:943–950

Scheffran J, Marmer E, Sow P (2012) Migration as a contribution to resilience and innovation in climate adaptation: social networks and co-development in Northwest Africa. Appl Geogr 33:119–127

Schmitt K, Albers T, Pham TT, Dinh SC (2013) Site-specific and integrated adaptation to climate change in the coastal mangrove one of Soc Trang Province, Viet Nam. J Coast Conserv 17:545–558

Scyphers SB, Picou JS, Brumbaugh RD, Powers SP (2014) Integrating societal perspectives and values for improved stewardship of a coastal ecosystem engineer. Ecol Soc 19(3):38

Serrao-Neumann S, Schuch G, Harman B, Crick F, Sano M, Sahin O, van Staden R, Baum S, Choy DL (2015) One human settlement: a transdisciplinary approach to climate change adaptation research. Futures 65:97–109

Shepard CC, Agostini VN, Gilmer B, Allen T, Stone J, Brooks W, Beck MW (2012) Assessing future risk: quantifying the effects of sea level rise on storm surge for the southern shores of Long Island, New York. Nat Hazards 60:727–745

Singh K, Walters BB, Ollerhead J (2007) Climate change, sea-level rise and the case for salt marsh restoration in the Bay of Fundy. Canada Environ J 35(2):71–84

Sultana P, Thompson P (2010) Local institutions for floodplain management in Bangladesh and the influence of the Flood Action Plan. Environ Hazards 9:26–42

Thornbush M, Golubchikov O, Bouzarovski S (2013) Sustainable cities targeted by combined mitigation–adaptation efforts forfuture-proofing. Sustain Cities Soc 9:1–9

Tompkins EL (2005) Planning for climate change in small islands: insights from national hurricane preparedness in the Cayman Islands. Global Environ Chang 15:139–149

Tribbia J, Moser SC (2008) More than information: what coastal managers need to plan for climate change. Environ Sci Policy 11:315–328

Uy N, Takeuchi Y, Shaw R (2011) Local adaptation for livelihood resilience in Albay, Philippines. Environ Hazards 10:139–153

Van Slobbe E, de Vriend HJ, Aarninkhof S, Lulofs K, de Vries M, Dircke P (2013) Building with Nature: in search of resilient storm surge protection strategies. Nat Hazards 66:1461–1480

Wardekker JA, de Jong A, Knoop JM, van der Sluijs JP (2010) Operationalising a resilience approach to adapting an urban delta to uncertain climate changes. Technol Forecast Soc 77:987–998

Woodruff DS (2010) Biogeography and conservation in Southeast Asia: how 2.7 million years of repeated environmental fluctuations affect today's patterns and the future of the remaining refugial-phase biodiversity. Biodivers Conserv 19:919–941

Wu H-C, Lindell MK, Prater CS, Samuelson CD (2014) Effects of track and threat information on judgments of hurricane strike probability. Risk Anal 34(6):1025–1039

Index

A
Acadian Peninsula (Shippagan), 31
Acadian Peninsula (Ste.-Marie-St.-Raphaël), 31
Action-oriented, 46
Actor perception(s), 45
Adaptation planning (workshops), 20, 29, 36, 45, 49
Adaptive capacity, 2, 11, 19, 29, 45, 47, 49, 68
Aged aging communities, 2, 10–12, 23, 35, 49

B
Baie des Chaleurs (Bonaventure), 31
Baie des Chaleurs (Maria), 31
Behavior behavioral change, 8, 22, 24, 47, 78, 80
Building permit(s), 81

C
Case study(ies), 3, 4, 19, 29, 31, 62, 66, 69, 82
Climate change adaptation, 3, 8, 21, 23, 30, 31, 36, 46, 56, 67, 68, 79, 85
Coastal Community Challenges-Community-University Research Alliance (CCC-CURA), 30, 31, 37, 39, 56, 59, 70, 72
Coastal defences protection, 44
Coastal development, 84
Coastal erosion, 21, 84
Coastal management, 68
Co-construct(ed) solutions, 3, 21, 30, 36, 69, 73
Collaboration, 21, 78, 86
Collective action, 21–23, 67
Collective capacity, 19, 21
Communication(s), 2, 12, 22, 45–49, 56, 78, 81, 85
Community action participation response awareness, 22
Community coherence, 13
Community solution(s), 21
Cooperation, 21, 49
Coping capacity, 7, 20
Cultural dimension, 23

D
Demographics, 5, 8, 12, 17, 23, 35, 78, 81

E
Education, 2, 10, 12, 20, 35, 45, 48, 49, 69, 78, 85
Emergency response(s) plan(s), 49
Emotional response(s) reaction(s), 5, 42
Engagement, Multistakeholder, 45
Erosion protection (measures), 21
Estuary (Ste.-Flavie), 31, 36
Evacuation, 12, 18, 42, 44, 48, 78, 80
Experience(s), 1–5, 8, 10, 12, 17, 21, 23, 24, 30, 32, 45–48, 55, 56, 77, 79, 81, 82

F
Floods flooding, 1, 2, 7, 11, 12, 32, 42, 44, 67, 78, 79, 82

G
Gender (mainstreaming), 17
Governance
 bottom-up, 19, 22, 23, 39, 46, 49, 71

Governance (*cont.*)
  good, 71, 79
  multilevel, 4, 49, 68
  sustainable, 23
  top-down, 23, 24, 39, 71
Government intervention(s), 20
Gulf (Rivière-au-Tonnerre), 31

H
Hazard-proofing, 44, 79
Hazard(s), 4, 10, 12, 21, 24, 29, 56, 68, 69, 81, 84

I
Impact(s), 2, 3, 8, 11, 12, 18, 21, 31, 35, 42, 49, 59, 62, 66, 69, 72, 78–80, 83
Indigenous people groups rights (social justice) information, 11, 45, 49, 85
Infrastructure(s), 1, 7, 10, 12, 20–22, 32, 42, 48, 50, 65, 68, 79, 85
Integrated adaptive management, 72
Integrated approach, 8, 44, 80, 83
Interdisciplinary (Interdisciplinarity), 3, 45, 86
Isolation (cut off), 66, 70

K
Knowledge co-production, 49

L
Local (governments), 20, 39
Local Service District (LSD), 4, 31, 50
Longitudinal, 3, 5, 17, 30, 45, 56, 62

M
Marginalized groups, 11
Meetings, 20, 36, 39, 78, 80, 85
Method of Evaluation by Group Facilitation (MEGF), 23, 24, 69
Migration-out, 10, 22
Migration relocation, 19
Multiscale (action) Multiscalar, 4
Multisite, 18
Municipalities(Municipal), 31, 36, 57, 70

N
New Brunswick (NB), 4, 31, 41
Northumberland Strait in the south of the Gulf (Cocagne Grande-Digue), 31
Northumberland Strait in the south of the Gulf (Dundas), 31

O
Occupation, 1, 8, 11, 12, 47, 68, 78, 81

P
Participatory Action Research (PAR), 3, 21, 22, 24, 30, 38, 39, 46, 69, 79
Participatory approach, 20, 23, 46, 72, 77, 83, 85
PEI (Morell), 31
PEI (Stratford), 31
Perception, Social, 3
Poverty, 8, 11, 13, 85
Preparation, 4, 18, 48
Preparedness, 12, 18, 35, 57, 60, 62
Prince Edward Island (PEI), 4, 8, 17, 42, 77
Proactive approach, 19, 46
Protection structures defences, 84
Psychosocial
  stressors, 45

Q
Québec(QC), 4, 31, 42, 56, 77

R
Reactive strategies, 20
Regional resilience, 61
Resource-dependent communities, 47, 85
Retreat strategy (ies), 84
Risk(s), 4, 7, 8, 11, 12, 18, 29, 30, 45, 47, 61, 70, 81, 84, 85

S
Social acceptance (acceptability), 39, 69
Social capital, 32, 60, 62, 85
Social cohesion (connectivity), 22, 60, 61
Social dimension, 56, 67, 85
Social-Ecological System (SES), 3, 4, 61, 62, 67
Social innovation, 83
Social learning, 12, 21, 46, 79, 80, 85
Social memory forgetting, 78
Social resilience, 35, 59, 67, 79, 80
Sociocultural dimension, 10
Sociopolitical, 68
Sociopscyhological effect(s), 24
Spatial planning, 45
Storm(s) Hurricanes, 7
Storm surge(s), 1, 7, 22, 41, 44, 47, 56, 79, 84
Storms, winter, 2
Sustainability
  environmental, 3, 4, 19, 29, 86

integrated approach socioeconomics, 19
Sustainable development, 29, 49, 71, 80

T
Tide(s), 22, 42, 56
Traditional knowledge, 45
Transformability, 68
Transformation, 46

V
Vulnerability, 2, 4, 7, 8, 11, 19, 22, 36, 47, 68, 73, 79, 85
Vulnerable communities, 12, 82, 84

W
Waves, 7, 11, 42, 43, 79
Webinars, 20

Made in the USA
San Bernardino, CA
01 September 2017